Whole Numbers Arithmetic: Essential Mathematics

Whole Numbers Arithmetic: Essential Mathematics

Editor: Brian Benson

MURPHY & MOORE
www.murphy-moorepublishing.com

Published by Murphy & Moore Publishing
1 Rockefeller Plaza,
New York City, NY 10020, USA
www.murphy-moorepublishing.com

Whole Numbers Arithmetic: Essential Mathematics
Edited by Brian Benson

International Standard Book Number: 978-1-63987-569-6 (Hardback)

Cataloging-in-Publication Data

Whole numbers arithmetic : essential mathematics / edited by Brian Benson.
p. cm.
Includes bibliographical references and index.
ISBN 978-1-63987-569-6
1. Numbers, Natural. 2. Numbers, Rational. 3. Arithmetic. 4. Mathematics.
I. Benson, Brian.
QA141 .W46 2022
513--dc23

Contents

Preface

Every book is initially just a concept; it takes months of research and hard work to give it the final shape in which the readers receive it. In its early stages, this book also went through rigorous reviewing. The notable contributions made by experts from across the globe were first molded into patterned chapters and then arranged in a sensibly sequential manner to bring out the best results.

Arithmetic is among the elementary branches of mathematics. It involves the study of numbers, relations between them, and observations on numbers. These are studied to solve various number problems. Addition, subtraction, division and multiplication are the basic arithmetic operations. Arithmetic generally refers to the elementary aspects of the theory of numbers, arts of measurement, and numerical computation. Counting is the act of determining the number of objects present in a set. The numbers thus obtained are called natural numbers. When no object is present, the count produces the number zero, which added to the natural numbers, comprises the set of whole numbers. This book is a valuable compilation of topics, ranging from the basic to the most complex subjects related to whole numbers arithmetic. Some of the diverse topics covered herein address the varied branches that fall under this category. For all those who are interested in whole numbers arithmetic, this book can prove to be an essential guide.

It has been my immense pleasure to be a part of this project and to contribute my years of learning in such a meaningful form. I would like to take this opportunity to thank all the people who have been associated with the completion of this book at any step.

Editor

1

Reflecting on the What and Why of Whole Number Arithmetic

Roger Howe

Introduction

Whole number arithmetic is a basic part of mathematics education everywhere, and there is a tendency, among mathematicians especially, but also I believe among mathematics educators, to think of it as 'easy' or 'simple', and in comparison to later parts of the curriculum – fractions, algebra, etc. – this attitude has some validity.

But, even WNA is not simple! At least, if history can be a guide to what is simple and what is hard, then it shows that our current, essentially universal, formu-lation of arithmetic in terms of base-ten place value notation (PVN) is not simple. It is the product of a very long historical development, some parts of which are still little understood. PVN has won its worldwide acceptance and use, not because it is simple, but for the power it brings to numeration and calculation; because once you understand it – or at least, can work with it – many other things are simple. Although PVN has that characteristic feature of a great idea – that after you learn it, you can't imagine not knowing it – it presents considerable obstacles to the learner, and many students master it only partially. PVN is not obvious. Chapter gives us some per-spective on why this is so.

Problematics of Place Value Notation

As we learn in Chap. the ancient or classical civilisations – Mesopotamian

R. Howe (✉)
Yale University, New Haven, CT, USA
e-mail: roger.howe@yale.edu

(Sumerian, Babylonian, etc.), Egyptian, Chinese, Indus Valley, Greek and Roman – although they used a lot of mathematics and did complicated calculations, none of them developed positional notation (although, in some sense, the Chinese came close). It seems to have come into being in South or Southeast Asia, perhaps not until the early centuries of the CE, as a result of interactions between Chinese merchants and various neighbours of China, more specifically as a result of trying to adapt Chinese computational methods to written form. The Chinese certainly had a highly developed base-ten arithmetic, and they already had many of the components of place value notation, especially as embedded in their computational tools, but their written numbers included explicit symbols for each base-ten unit. They had no symbol for zero; if the multiple of a particular unit needed to express a given number was zero, the unit was just omitted from the writing. The Romans used a base-ten system, but their notation was more or less what is designated in Sect. as a system of 'additive type', with no digits. They had separate symbols for each base- ten unit and also one for five of each unit, and they used subtraction as well as addition to express numbers. Although we think of the classical Greeks as strong mathematicians and they did marvellous things in geometry, their conception of number was not as advanced. The Greek numeration system was quite ad hoc and limited. It used combinations of up to three letters to express numbers up to 999. (As related in a draft version of Chap., the Hebrew and early Arabic systems were similar – was this perhaps a common legacy from Phoenician practice, as with their alphabets themselves?) So the late development of place value notation, along with the gradual evolution of the computational algorithms after it was invented, provides convincing historical evidence that PVN is far from obvious.

Carl Friedrich Gauss (1777–1855) is often named as the best post-Renaissance European mathematician. Although pure mathematicians today revere him especially for his contributions to number theory, he also was an accomplished applied mathematician. He did massive calculations by hand, including a determination of the orbital motion of the large asteroid/dwarf planet *Ceres*, from only a very few observations of its position. His prediction of where in the sky to find it again led to its rediscovery and its establishment as a notable member of the solar system.

Gauss's experiences with computation made him highly sensitive to the virtues of place value notation. He is quoted (Eves 1969) (Newman 1956, p. 328) as saying:

> The greatest calamity in the history of science
> was the failure of Archimedes
> to invent positional notation.

Here, the main point is that even Archimedes, perhaps the best of classical Greek mathematicians, had not come up with the idea of place value notation. This was not because Archimedes never thought about large numbers: he wrote a paper called *The Sand Reckoner*, whose theme was precisely how to express large numbers (such as the number of grains of sand in all the beaches of the world), and which discussed ways of constructing and naming such numbers. Nevertheless, the concept of an efficient, all-purpose, unlimited system for writing numbers, such as is provided by place value notation, did not figure in his proposals. One might wonder whether the very limited ad hoc system (described above) in common use by classical Greeks and, in particular, its lack of use of multiplication, somehow inhibited Archimedes' thinking.

Algebraic Structure and the Power of Place Value Notation

If PVN was so highly valued even by Gauss, it is worth looking at carefully. As noted in Sect. , positional notation harnesses many concepts, including essen-tially all of basic polynomial algebra, in the service of simply writing numbers. Moreover, it owes it computational efficacy to its compatibility with algebraic structure. It also relies on numerous conventions that must be understood by the reader in order to interpret base-ten numbers correctly. Three different interpretations or elaborations of place value notation are mentioned in Sect. of Chap. as having been taught in France in recent years (in addition to the 'academic theory' promoted by the New Math). My paper for the Study Conference (Howe 2015) describes a slightly more elaborated sequence of interpretations, five in all, that includes all of these in a single developmental scheme that might be taken as five stages of place value understanding and that could describe the progression of a learner through a carefully designed arithmetic curriculum. They are illustrated by the following sequence of equations:

$$\begin{aligned} 456 &= 400 &&+ 50 &&+ 6 \\ &= 4 \times 100 &&+ 5 \times 10 &&+ 6 \times 1 \\ &= 4 \times (10 \times 10) &&+ 5 \times 10 &&+ 6 \times 1 \\ &= 4 \times 10^2 &&+ 5 \times 10^1 &&+ 6 \times 10^0 \end{aligned}$$

The first interpretation of '456' mentioned in Sect. is more or less the way '456' is read out loud in English. (Apparently, the practice is different in France, with the base-ten units suppressed.) The second and third interpretations agree with the second and third stages above.

The second stage breaks the number up into a sum of pieces. In terms of PVN, each of these pieces involves only one non-zero digit. This stage displays the basic strategy of expansion in a given base: every number can be expressed as a *sum* of pieces of a special kind. Remarkably, there appears to be no standard short name for these pieces in the mathematics or mathematics education literature. For present purposes, we will refer to them as *base-ten pieces*.

Thus, the base-ten pieces of 456 are 400, 50 and 6. This explicit decomposition of a number into the sum of its base-ten pieces is often presented in US classrooms under the name *expanded form.*

The base-ten pieces of a number themselves have substantial structure, which are to be understood in terms of multiplication. The third to fifth stages of PVN reveal successive features of this multiplicative structure.

The first aspect of this structure is that each base-ten piece is a multiple of an even more special quantity, a *base-ten unit.* These are the base-ten pieces whose non-zero digit is just 1. A general base-ten piece is a multiple of a base-ten unit, and the non-zero digit tells us what that multiple is. Thus, $400 = 4 \times 100$, $50 = 5 \times 10$ and $6 = 6 \times 1$. Thus, the base-ten units here are 100, 10 and 1.

With this terminology, we can say:

Each base-ten piece is a digit times a base-ten unit.

This is what the third stage is telling us. It might be called the 'second expanded form'.

We should not stop with this stage. The base-ten units themselves have multiplicative structure, and this structure is really the key to the efficacy of the idea of base-ten expansion, so it should be made explicit.

Each base-ten unit is itself a product: a repeated product of 10s. The base-ten unit 1 indicates the basic unit of the whole system. It stands for whatever quantity you are counting. As revealed already by prehistoric tally systems, all whole numbers are obtained by iterating the basic unit sufficiently many times. The next base-ten unit, 10, is the key to the whole system. It reveals the *base* or grouping ratio: each successive base-ten unit is obtained by combining 10 of the previous unit. Thus, $10 = 10 \times 1$ is the first base-ten unit beyond unity. The next base-ten unit is $10 \times 10 = 100$. The next one is $10 \times 100 = 1000$, and on and on, for as long as we need to go. For most everyday purposes, we don't have to go too far: since each base-ten unit is ten times the next smaller one, these numbers get large fast! In our example, we only need the first three units – 1, 10 and 100. The standard Greek, Hebrew and Arabic notational system was content with representing numbers only up to 999. Roman numerals went somewhat farther but not much.

Finally, the last expression summarises the previous one by expressing the iterated products of 10 with itself in exponential notation. It is the 'academic theory' specialised to base ten. The expression bears strong resemblance to the way polynomials are written in algebra and indeed can be thought of as expressing the given number as a 'polynomial in 10', with the understanding that the 'coefficients' of the 'polynomial', i.e. the digits in the first expression, are all whole numbers less than 10.

From an educational perspective, a key point to realise about the five stages of place value is that although to a mature understanding all these expressions are more or less obviously equivalent, each stage represents a substantial intellectual advance on the previous one. For example, exponential notation is justified using the associative rule for multiplication, which is arguably the deepest of the rules of arithmetic. Correspondingly, exponential notation is usually not introduced until late elementary school – not in the primary grades – and well after PVN has already been in use by students with 3–5-digit (or more) numbers. (However, this late introduction is probably not done to allow a principled discussion of the role of the associative rule in defining exponential notation!) Taken all together, understanding the five stages implies a lengthy intellectual development that requires, if it is achieved at all, the full span of elementary education.

Possible Lessons for Education

All this is well known to mathematics educators, but it seems worthwhile to rehearse it again here, for several reasons.

First, it can help us to see to what may be some omissions in educational discourse and in WNA instruction. The first example of this would be the lack of a short name for talking about the basic building blocks of the notation, what we are here calling the 'base-ten pieces'. A short name would facilitate discussion of them and their role in PVN, thereby promoting conceptual understanding of PVN.

Further on, we should ask to what extent the full structure revealed in the fifth expression, that is, the algebraic structure implicit in PVN, is made clear to students. Some evidence suggests that, in the United States, even the third stage, the 'second expanded form', does not become part of the thinking of a large segment, perhaps a majority, of students (Thanheiser 2009, 2010). We should consider how to structure our curricula so that the ideas embodied in the five stages of place value, and the algebraic structure underlying it, are absorbed by students. This would be consistent with the 'higher-order thinking' mantra of twenty-first century education.

Second, the unlimited nature of place value notation is brought home when we use exponential notation to express base-ten pieces as $d \times 10^k$, where we understand that k can be any whole number. The WNA curriculum as we have inherited it from previous centuries might be described as 'small-number-centric' (which was appropriate for many of its primary justifications, e.g. 'shopkeeper arithmetic'). It starts with single digits, proceeds to two-digit numbers, then to three- and four-digit numbers, and then perhaps with a little attention to five- and six-digit numbers, tends to think its job is done. This is probably fine if the main goal of instruction is to enable people to calculate correctly with medium-sized numbers, but it does not convey a sense of the system as a whole. This small-number focus may be part of the reason many people have little appreciation for the difference between a million and a billion, thinking of both of them simply as 'very big numbers'.

But one can argue that today, it is an important civic skill to understand the difference between a million and a billion. To understand, for example, that a billionaire is equivalent to a thousand millionaires.[1] If a billionaire spends 10 million dollars on a house and takes a million dollar vacation each year, he still has about $950 million left to do other things with. When Bill Gates built his house, everyone

[1] This is in the USA. In England, 'billion' means 'million million'.

was agog that he spent 40 million dollars on it. But Gates' net worth at the time was 40 billion dollars, so he was spending 0.1% of his net worth on the house. What kind of a house could you buy for 0.1% of your net worth? In this context, it should also be taken into account that rich people do not keep their money under their pillows, they invest it: their money makes more money. If Gates' fortune was increasing at just 1% per year (in fact, it was increasing much faster), he was richer after he paid for the house than when he signed the contract to have it built.

Furthermore, to discuss intelligently economic constructs such as gross national product, billions are not enough; one needs at least trillions. For example, the GDP of the United States in 2015 was about 18 trillion dollars.

Discussion of issues like climate change involves comparing numbers that are substantially larger than this. For example, how big is Earth's atmosphere? The weight of the atmosphere is about 15 pounds per square inch (atmospheric pressure) times the area of Earth in square inches. How many pounds is that? Computations like this one also bring home the point that, not only would it be very cumbersome to attempt to compute all the digits in a large number, it would be a waste of time.

With some effort, one can compute by hand that the number of square inches in a square mile is 4,014,489,600. (This computation does not exceed the capacity of many hand calculators, so a modern child can find it with a few keystrokes – if she/he knows what to do.) Then all we have to do is multiply this by the area of Earth in square miles. But how accurately can we know this? Do we want to try to find the exact true area of Earth in all its glorious roughness? This does not even make sense. Most of the surface of Earth is water, which is constantly being jostled by the wind to form waves. Waves change the surface area of water, sometimes quite drastically. (Think of 'The Great Wave off Kanagawa' in Hokusai's woodblock print. Some would put this forward as an example of a fractal, with infinite surface area.) A simpler approach might be to pretend that Earth is a sphere and use the formula $A = 4\pi r^2$ for its area. To carry out this strategy, you have to confront the facts that (i) Earth is not in fact a sphere and, in particular, (ii) its 'radius' is not exactly defined. In fact, the 'radius of Earth' does not make sense to much more accuracy than ±5 miles.[2] Since it is approximately 4000 miles, this means that we know the radius of Earth to less than three significant figures. Keeping in mind the principle that a product is only known as accurately as the least accurate of its factors, it does not make sense to report more than the three largest base-ten pieces in describing the area of Earth or the weight of the atmosphere. So our lovely calculation above of the number of square inches in a square mile could (and should) just be replaced by 'approximately 4 billion'. The corresponding figure for the area of Earth in square inches is quite adequately represented by 800,000,000,000,000,000 or 800 quadrillion (in US numeration).

To successfully teach children to comprehend and work with numbers this large, we have to get away from focusing on the digits and work hard to understand the sizes of the pieces: to focus more attention on the base-ten pieces, especially the

[2]There are several reasons for this: oblateness (flattening at the poles), the bulge in the North Pacific, mountains and ocean trenches, etc.

base-ten units and their relative sizes. Furthermore, students need to learn that, as in the example of 'radius of Earth', in the real world, it is rare either to need to know, or even to be able to know, more than the two or three largest base-ten pieces in the number describing some quantity. For most practical purposes, a number is a two- or three-digit number times a (perhaps large) power of 10. This might be a goal for WNA in the twenty-first century.

It is tempting to speculate why the New Math reform of the 1960s did not articulate the process of learning place value as completely as the five-stage description offered above. Perhaps it was a lack of pedagogical insight on the part of participating mathematicians or, more precisely, failure to appreciate the intellectual advances and the years of development needed to progress from one stage to the next. Perhaps it was a residue of the somewhat contemptuous attitude some mathematicians harbour towards base-ten notation, since it involves arbitrary choices and, especially, selection of a base, for which there is no clear mathematical reason. This might explain the introduction of arbitrary bases in the New Math. Perhaps it was because they were still so starry-eyed about the triumph of set theory in establishing foundations for mathematics that mundane classroom issues such as arithmetic did not engage their attention. Whatever the reason(s) behind it, this failure can serve as an exhibit for the claim that mathematical expertise is not the only prerequisite for understanding and positively influencing mathematics education.

Comments on Particular Sections of other Chapter

Comments on all Section

The linguistic issues of teaching the base-ten place value system are perhaps the feature of WNA that benefits most from cross-national comparisons. In the USA, we have been aware since the paper (Miller and Zhu 1991) pointing out the strict compatibility of Chinese spoken number names with PVN and the comparative disadvantage English speakers have in learning the principles of place value, since they are obscured at the beginning by the irregularity of the -teen numbers and, to a somewhat lesser extent, of the -ty numbers. However, we learn in Section that English presents relatively mild problems and that several European languages such as French and Danish are much worse in this regard. My heart goes out to Danish school children trying to make sense of 70 when the name for it is 'three and a half four'!

[3]The section numbers refer to sections of Chap.

In Sect., we learn that children in Algeria have the added burden of trying to translate between several different vernaculars with contradictory conventions, compounded by translational ambiguities.

Perhaps a way to help children get around this kind of linguistic obstacle is to treat the base-ten system for what it is in almost all countries – an imported piece of a foreign language – and to make the translation from traditional number names to 'structural names', or 'mathematics names', that explicitly describe the base-ten structure of each number, a topic of study. This would include explicitly discussing the -teen numbers as being made from one 10 and some 1s and making sure that students could translate between their traditional names and the structural descriptions. Likewise, the -ty numbers (20, 30, …, 90) would be explicitly identified as a certain number of tens and the general two digit numbers as being the sum of some 10s and some 1s. The work of Fuson (e.g. Fuson and Briars (1990), Fuson et al. (1997)) gives some support to this approach.

Beyond helping children to translate between their traditional names and the quantity meanings of numbers with two or three digits, this approach would have the advantage of permitting explicit attention to be paid to the base-ten pieces and how the base-ten structure facilitates computation. General rules could be enunciated that describe what needs to be done to add or to multiply.

> When adding two base 10 numbers, we add the 1s (from the two numbers) together, we add the 10s together, we add the 100s together, etc. Then, if we get more than 10 of any base 10 unit, we convert 10 of it to 1 of the next higher unit.
>
> When multiplying two numbers, we multiply each base 10 piece of one factor, with each base 10 piece of the other factor. Then we sum all these products.
>
> The multiplication of two base 10 pieces amounts to multiplying their digits, and multiplying the base 10 units, and taking the product of these.

These descriptions can be stated briefly or more completely, as appropriate for the context. Besides encapsulating the main principles of base-ten computation in compact form, these general rules have the advantage that they can be taught progressively, starting with two-digit addition and one-digit by two-digit multiplication, and can be formulated more and more generally as students work with larger numbers. The standard column-wise procedures for paper-and-pencil calculation can then be presented as mechanically simple ways to actualise the principles of addition and multiplication formulated as above. This approach would also easily afford discussion of the role of the rules of arithmetic in ensuring that the general recipes formulated above are valid. All this would in turn prepare students for learning the later stages of place value, in the list of five stages given above, and give them a chance of understanding the whole system, which currently seems to be a rare accomplishment.

The misunderstanding, between the university mathematician Rick and several K-12 mathematics teachers, mentioned by Cooper (2015), makes an interesting study for another mathematician. The miscommunication may have occurred in the interpretation of 'numerical expression'. Rick may have meant this to mean, 'Does this symbol "8(1)" signify a number?', while the teachers may have taken it to mean, 'Is this a well-defined expression involving numbers?' It is the latter, but it is not the former.

The situation is primed for confusion by the frequent use of the word 'division' to signify either division in the usual sense of rational numbers or division with remainder (DWR).

DWR is not an operation on whole numbers in the same sense that addition or multiplication is an operation. That is to say, DWR does not take a pair of whole numbers and return a single whole number: it produces a *pair* of whole numbers, which play very different roles in the process of DWR. One number is the DWR 'quotient' and the other number is the remainder. Considered in this way, DWR defines a somewhat complicated function from pairs of whole numbers to pairs of whole numbers. It is a rather different animal from rational number division, which takes a pair of rational numbers and returns a *single* rational number.

The confusion is further encouraged by the use of the notation 25:3, which is very similar to the usual fraction notation 25/3 (and probably intended to be similar!). The symbol 25/3 denotes the (rational) number x such that $3x = 25$. However, the notation 25:3 stands for the pair of whole numbers q and r, such that $3q + r = 25$, with r understood to satisfy $0 \leq r < 3$. These numbers are of course 8 and 1.

To emphasise the difference between DWR and the operation of division for rational numbers, instead of writing 25:3 = 8(1), which is trying to make DWR look as much as possible like actual rational number division (RND), we might try to emphasise the distinction and try to make DWR look different from RND. To do this, we might define the 'DWR function', which would take a pair of whole numbers (n,d) to the pair of numbers (q,r) such that $n = qd + r$, with the remainder r understood to satisfy the key condition $0 \leq r < d$. Thus, we would write

$$\text{DWR }(25,3) = (8,1)$$

to emphasise the function aspect of DWR. (This notation might not be so student-friendly, however!) The DWR function is not one-to-one (in fact, it is infinite-to-one), is not RND and indeed is not compatible with RND, or with multiplication in the whole numbers, so there should not be any expectation that, just because

$$\text{DWR }(25,3) = (8,1) = \text{DWR }(33,4) = \text{DWR }(41,5) = \text{DWR }(49,6) = \text{DWR }(57,7),$$

etc., that we can conclude that 25/3 = 41/5 = 57/7, etc. But the notation DWR(25,3) = 25:3 tries to make DWR look like RND and sets things up for the confusion that Rick and his teachers experienced.

Besides the definition of the DWR relation as deriving from the equation $n = qd + r$, there is another conventional notation that can adequately express the relationships involved in DWR – the notation of mixed numbers. This would allow us to write

$$\frac{25}{3} = 8\frac{1}{3}$$

Here the conventional interpretation of the right hand side is as a sum:

$$8\frac{1}{3} = 8 + \frac{1}{3}$$

This notation is more or less equivalent to the notation that Rick proposed to the teachers, but avoids the tricky sign, whose interpretation is at the core of the confusion. Understanding that 25:3 = 8(1) means that $\frac{25}{3} = 8\frac{1}{3}$ and that 41:5 = 8(1) means that $\frac{41}{5} = 8\frac{1}{5}$ should help cure people from wanting to conclude equality of the left-hand sides implies equality of the right-hand sides. The main point to be clear on is that, while it might be defensible to call 8(1) a 'numerical expression', it is *not* a number.

Conclusion

In these comments, I have tried to reinforce the theme of Chap. that WNA is not a simple matter. This can be seen in the historical development of the base-ten place value system, which was at best incompletely realised by early civilisations that began using mathematics heavily. It can also be seen in the conceptual structure, which is at best rather incompletely taught in many countries. Finding better, more effective and more conceptually complete ways of teaching WNA should be a focus of research. Finally, the fact that many issues of current importance (national budgets, climate science, big data) require dealing with large numbers that are known only approximately indicates placing more instructional emphasis on a global understanding of base-ten structure and, in particular, on the base-ten pieces and their relative sizes.

References

Eves, H. (1969). *In mathematical circles*. Boston: Prindle, Weber & Schmidt.
Fuson, K. C., & Briars, D. J. (1990). Using a base-ten blocks learning/teaching approach for first

and second grade place-value and multidigit addition and subtraction. *Journal for Research in Mathematics Education, 21*(3), 180–206.

Fuson, K. C., Smith, S. T., & Lo Cicero, A. (1997). Supporting Latino first graders' ten-structured thinking in urban classrooms. *Journal for Research in Mathematics Education, 28*(6), 738–766.

Miller, K. F., & Zhu, J. (1991). The trouble with teens: Accessing the structure of number names. *Journal of Memory and Language, 30*(1), 48–68.

Newman, J. R. (1956). *The world of mathematics*. New York: Simon and Schuster.

Thanheiser, E. (2009). Preservice elementary school teachers' conceptions of multi-digit whole numbers. *Journal for Research in Mathematics Education, 40*(3), 251–281.

Thanheiser, E. (2010). Investigating further preservice teachers' conceptions of multi-digit whole numbers: Refining a framework. *Educational Studies in Mathematics, 75*(3), 241–251.

Cooper, J. (2015). Combining mathematical and educational perspectives in professional development (pp. 68–75).

Howe, R. (2015). The most important thing for your child to learn about arithmetic (pp. 107–114).

2

How to Teach and Assess Whole Number Arithmetic

Claire Margolinas

Preliminary Considerations About Teaching and Assessing

Teachers are accountable for classroom interactions and pupils' work assessment. However, those visible actions (for instance, during the lesson observed by working group 4 participants in Macao) are only a part of teachers' work, and include also:

- *Planning* not only a single lesson but also a sequence of lessons and more generally thinking about and designing the entire mathematical theme (for instance, WNA as a whole). This usually depends on use of resources that are selected by the teacher.
- *Selecting* the physical objects to be used during the lesson (or the sequence of lessons), the textbook which can be used by the pupils and/or by the teacher as a source of inspiration, the tasks that might be designed by others and are available (by sharing with colleagues, by browsing the Internet, etc.), the items for the assessment, etc.

These aspects of teaching require teacher knowledge, which is not easily observed, since it is accessible only by means of what the teacher might say about her activity, which is always a reconstruction on her part, and what the teacher is doing in the classroom, which is subject to diverse interpretations.

Various aspects of teacher knowledge have been considered within different frameworks that all take into account *pedagogical content knowledge* (Shulman 1986). This model has been refined by Deborah Ball and her colleagues (Ball et al. 2008), who have examined the impact *mathematical knowledge for teaching* on the quality of instruction (Hill et al. 2008). It is thus my purpose to highlight some

C. Margolinas (✉)
Laboratoire ACTé, Université Clermont-Auvergne, Clermont-Ferrand, France
e-mail: claire.margolinas@uca.fr

aspects of whole number arithmetic knowledge for teaching which seem important to promote interest-dense situations (Bikner-Ahsbahs et al. 2014) and thus the development of pupils' metacognitive strategies.

Whole Number Arithmetic and Mathematical Knowledge for Teaching

Chapter and working group 4 of the conference refer to 'teaching and assessing whole number arithmetic'; this formulation leads to consider WNA as a homogeneous domain. However, some very different aspects have been investigated in the conference papers:

- Understanding of numbers
- Place value and written numbers
- Understanding of operations
- Written operations (standard and non-standard)
- Memorisation of numerical facts: additive or multiplicative

In the book, the main topic has been the one chosen for the observed lesson: understanding of addition with carrying.

However, making coherent choices about WNA teaching requires organised knowledge of WNA (Askew 2015) and specific knowledge of concepts (see Barry et al. 2015 for a study about additive problems):

- Which sub-matters are related?
- How these sub-matters are linked together?
- What are the vital choices for teaching numbers?

In order to approach those questions related to the teaching and assessing of WNA, I will take three examples and then return to the Macao lesson.

Memorising Numerical Facts

During early primary school, pupils are engaged in various memorisation activities: they memorise nursery rhymes and poems, the number name file, the days of the week, the name of the months, the names of their friends, etc. Some years later, they will memorise a lot of facts and rules: grammatical, historical, mathematical rules and facts, etc.

What specific knowledge do teachers need in order to help pupils to memorise numerical facts? Is this different from the memorisation of nursery rhymes? What about grammatical rules? Are all numerical facts alike in terms of memorisation?

Let's begin with the first numerical memorisation: oral number names. This sequence of words share some properties with all songs and rhymes: some parts are made from words without links (one, two, like Humpty Dumpty); some parts are similar (twenty-one is similar to thirty-one, like a chorus) and you have to say it in the right order. However, number names are special: because of their use for counting, in particular, the words have to be clearly separated (one/two/three and not onetwothree), and the exact words and order of the words are crucial. If the oral number file is a base-ten language (a lot of languages across the world are in base twenty, also known as vigesimal[1]), you have thus at least ten different terms to memorise. Those terms are not different from any other list of terms (days of the week, song, etc.). The following names depend on the language you use (see this volume, Chap.). If you are very lucky, you may live in a country where the oral numeration is regular: ten-one, ten-two, etc. If you are not so lucky, you will have to memorise other terms, for instance, eleven and twelve in English (and up to the name for 16 in French, etc.). The rest of the list will have some regularity and irregularity, thirteen instead of third-ten, for instance, or a vigesimal system at some point (for instance, from 60 up to 99 in France). Teachers have to be aware of nuances of language in order to understand when they have to treat the number name file exactly like a song and when they might help the children understand how it is built. *This is clearly mathematical knowledge for teaching: it is not mathematical common knowledge.* For instance, a majority of persons in France are not aware of a base twenty (it is different in Belgium, Canada or Switzerland), but for teachers it is vital knowledge to understand that it is quite strange to say *soixante-dix* (sixty-ten), but if you do, it is normal to proceed and say *soixante-et-onze* (sixty-and-eleven).

If we now take the memorisation of numerical facts about addition and subtraction, it is first to be noted that it is not obvious to know when you really enter 'addition'. For instance, you have to teach very early that if you say six after five in the oral number file, in consequence if you have five objects and another one, you will have six in all. This will be related later to $5 + 1 = 6$. However, it means that those additive facts (+1) are learned in a totally different way as, for instance, $5 + 3 = 8$.

If we proceed on our reflection about the memorisation of additive facts, there are different ways to give the answer promptly (see Cao et al. 2015 about memorisation of multiplication table). The first is to memorise every additive fact. For numbers less that ten, you have 10×10 facts to memorise, but we have already stated that perhaps you may not have to memorise +1 as an additive fact, which takes off ten results to memorise. If you know $6 + 1 = 7$, do you have to learn that $1 + 6 = 7$, or do you have to know that you can exchange the numbers in addition (commutativity)? There is a balance to be found between the memorisation of facts and the memorisation of properties (which is the very first element of algebraic thinking; see Wong et al. 2015 for other algebraic problems).

If we return to $6 + 7$, you may have memorised the answer among the 90 or 45 facts that remain in your list, or you can think that 'six and four, ten and three: thirteen'. In order to obtain 13 very rapidly, you should memorise the ten comple-

[1] https://en.wikipedia.org/wiki/Vigesimal

ments (five facts only if you consider commutativity) and the procedure to count with ten or multiples of ten. There is, thus, also a balance to be found between the memorisation of facts and the memorisation of procedure.

Furthermore, the choice of procedures is crucial. For the same addition, if you have learnt to calculate using 5 as a step, you will think 'five and five and one and two: thirteen' and not 'six and four and ten and three'.

Those mathematics reflections have a big impact on the teaching and assessing of additive facts. For instance, if you want to induce 5 as a step, you will teach all the decomposition of numbers using 5 as a step from the beginning of the teaching of number: 6 will be considered as 5 and 1, 7 as 5 and 2, etc., and those relationships to 5 will be memorised. To assess the memorisation of additive results will be considered with those results and subsequent procedures in mind. For instance, it would be considered as a really basic task to give the answer to 12 + 13, but more difficult to give the answer to 16 + 14 and even more 17 + 18. On the other hand, if you have memorised additive results using 10 as a step, 16 + 14 should be the easiest.

The role of researchers in mathematics education may have a great impact on teaching and assessing if they help teachers to understand how special mathematics considerations will impact their decisions when they plan their teaching and select their materials and also highlight the different aspects of 'memorisation'.

Writing Numbers and Numerical Sentences

Some aspects of mathematical writing are specific, and some are shared with all writing language experiences (see Sensevy et al. 2015 for a design which is based on writing mathematics). The aspect which is present in both cases is the possibility that writing offers to avoid painstaking memorisation of facts. Writing is thus always in concurrence with oral memorisation. Another common aspect of writing is the possibility to communicate to others with a spatial or temporal gap. Yet another is the bureaucratic function of writing: when you write, you can organise the objects, for instance, in columns and lines, in ways you cannot reproduce orally (Goody 1986). What is totally different is the lack of connection between sounds and writing: 216 is not read two-one-six, etc. Furthermore, 21 is read twenty-one but when you read 216 you do not hear twenty-one (it is not easy to understand that there are 21 somethings in this number: 21 tens). Another notable disparity is that between the quite universal understanding of written numbers and that of mathematical writing in general. Thus the use of writing in mathematics should be a specific part of teaching mathematics (and not only written standard algorithms; see Zhao et al. 2015).

For instance, suppose that you want to associate eggs with egg cups with a temporal gap (see Alafaleq et al. 2015 for equality problems in textbooks). You have the eggs one day and the egg cups the day after. It is difficult to memorise the number of eggs, and you might use writing for this task. When you implement this situation for pupils (5–6 years old), the use of numerical writing is required in this situation.

Depending on the number of eggs and pupils' knowledge of writing numbers with digits, they may struggle to find a suitable way to use writing.

Some pupils will try to draw eggs, using the right colour and the right shape, but not the right quantity. They will realise that their writing does not give any information when confronted with the egg cups. A successful procedure might be to use the spatial organisation of the eggs and try to draw a 'map' with the places of the eggs, drawing for instance round shapes. Another procedure is to draw little straight lines: one for each egg. Teachers should consider this procedure as a very interesting attempt at symbolisation and thus encourage this behaviour and not only consider writing with digits. There are different ways to write quantities, and the efficacy of writing depends on the situations you are dealing: in particular situations, even an adult might write ~~IIII~~ ~~IIII~~ ~~IIII~~ in order to keep a record of 15 objects.

It is not enough to acknowledge that writing mathematics is an important part of whole number arithmetic. How this process is approached will vary according to the teacher's interpretation, whether they see writing as a fixed set of rules or as a way to think mathematically. I will illustrate this with an example (Laparra and Margolinas 2009). During a session observed in class 1, pupils were asked to solve the following problem 'There are 12 squares in a box. There are red and blue squares. There are 5 red squares. How many blue squares?' Pupils had some difficulties to solve this problem: they had not studied subtraction previously, and it was the first time they had to solve a word problem. At some point during the lesson, when all pupils were convinced that 7 blue squares was the solution, the teacher asked them to write or draw something in order to explain their solutions. Hamdi (Fig.) had drawn 12 squares and crossed 5 squares: there are 7 non-crossed squares. The representation of the problem is particularly accurate.

If you read the mathematical sentence, 12 + 5 = 7, you might think that Hamdi has made a big error (that is what the teacher thought), but it is highly improbable that Hamdi thought twelve and five are seven. In writing this wrong sentence, Hamdi demonstrates his current knowledge of written operation. He has written the numbers in the order of the given problem (12, 5, 7) and also in the order he has made use of those numbers in the schema, which is wrong for mathematical sentences but right for linguistic sentences, although the sentence is mathematically well formed. Hamdi certainly knows that the result of calculation is normally after the equality sign, but he doesn't know how to combine the only signs he had already learned (+ and =) in order to explain his reasoning.

On the other hand, if you consider Floriane's production (Fig.), it is difficult to understand Floriane's solving procedure, but it is very interesting to discover a kind of prefiguration of an equation: $5 + x = 12$.

Fig. Hamdi's representation of the squares problem

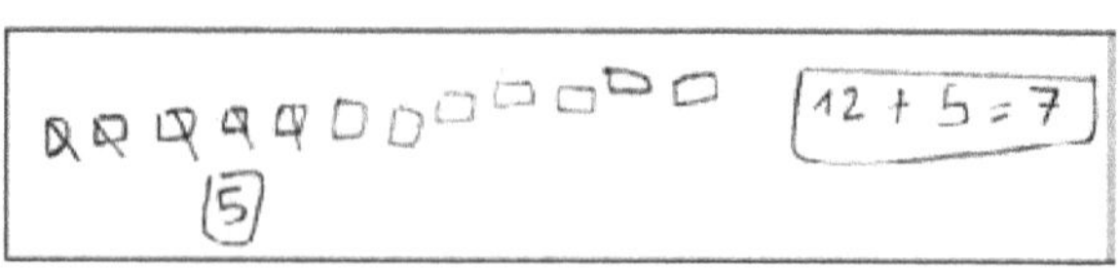

Fig. Floriane's representation of the squares problem

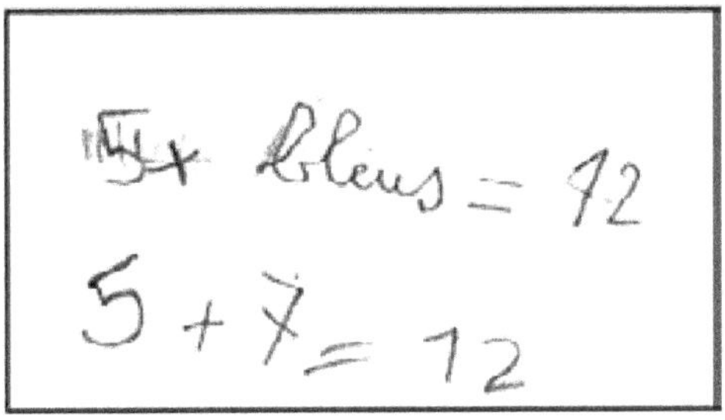

Both productions have their own qualities, and they reveal the difficulty to assess the production of written mathematical sentences. Unfortunately, the teacher was only interested in the correctness of the written addition.

There is certainly an important need for collective work in mathematics education in order to convey coherent knowledge about writing numbers and numerical sentences, since this is crucial for teaching and asserting WNA in general, at all levels of teaching.

The Field of Additive Structures

The expression 'field of additive structures' is taken from Vergnaud (1983, p. 31), whose work is of paramount importance in order to understand together addition and subtraction (for Chinese tradition).

The first comment based on Vergnaud's work is about subtraction – comparison between quantities is only one meaning for subtraction:

> The very first conception of subtraction for a young child is a "decrease" of some initial quantity [...]

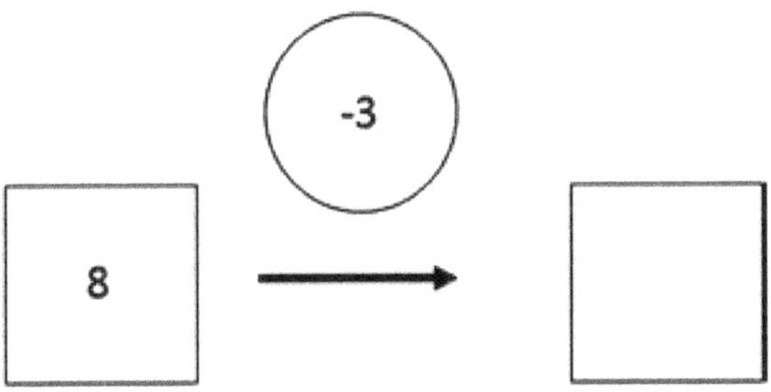

> *Example 1: John had 5 sweets, he eats 3 of them. How many sweets does he have now?*
> It is not straighforward, with such a conception in mind, to understand –subtraction as a relation of complements.

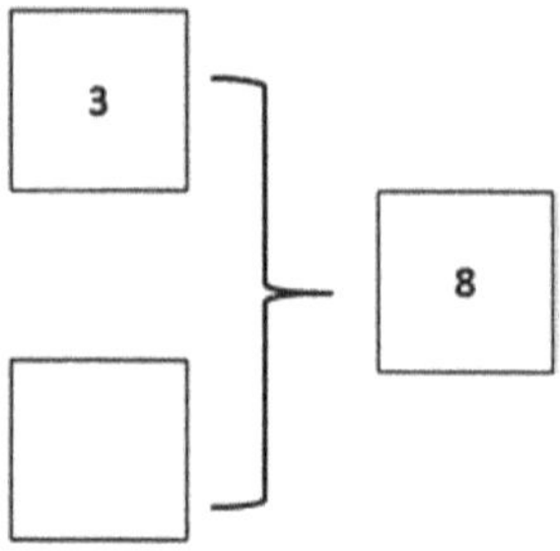

Example 2: There are 8 children around the table for Dorothy's birthday. 3 of them are girls. How many boys are there? (Vergnaud 1983, pp. 31–32)

In the following pages of the paper, Vergnaud enumerates and exemplifies the other conceptions of subtraction, as the inverse of an increase and as a difference relationship between states, between compared quantities and between transformations, and he concludes with: 'One can easily imagine the difficulties that children may meet in extending the meaning of subtraction from their primitive conception of a 'decrease' to all these different cases' (p. 32). Vergnaud has shown that pupils are able to solve the first problems from a young age, but the more difficult ones only at the end of primary school (even if the calculation, 8–3, remains the same). The same kind of differences of conception exists also for addition, both operations being regrouped in the additive structure field.

Those distinctions are essential for teacher mathematical knowledge, since they have to be aware of the nature of the problems which are proposed, in order to teach or assess addition and subtraction. The predominance of comparison has to be questioned.

The Macao Lesson: A Commentary

In this last part of this paper, I will give an overview of the points of the Macao lesson on addition, using the different aspects I have introduced earlier.

My first remark is about the subject matter of the lesson: it is well known that addition, even with carrying, is very easy, compared with subtraction (see Pearn 2015), and I would have been very interested to know how the skilful teachers of this school would have taught this challenging subject matter.[2] Even if I understand that the choice was not due to the working group, it is important to reflect on the most favourable environment of a scientific discussion about teaching and assessing WNA.

[2] Another group of participants (from other working groups) actually observed a lesson on subtraction in another school.

However, the lesson is very interesting in itself, and it is in a certain sense a model of mastery of a kind of lesson which can be observed also in other countries:

- The lesson begins with the oral recollection of numerical facts (number combination for ten), during which students are encouraged to give answers very rapidly.
- The teacher has introduced a material (candies and boxes of candies), pupils are mostly working with the material, they express their ideas orally, and they write with the scaffolding of the teacher.
- Different problems of addition of increasing difficulties are introduced with both representation of material and mathematical sentences (40 + 3; 25 + 2; 25 + 20) before the core topic of the lesson, which is to study an addition with carrying (24 + 9).
- The teacher is aware of a variety of possible answers for this problem, she has determined three procedures, and those procedures are represented by numerical sentences which are written in advance and ready to show to the pupils.
- The 'making-ten' strategy is clearly emphasised at the end of the lesson during the 'lesson summary'.

The calculations (Stage 1) in the oral phase represent nearly all the combinations for ten (only 0 + 10 is missing); thus, the answer is always ten. The first five questions are given in order (one number in the addition increases by one at every step). Thus, this first part of the lesson can be considered as a systematic presentation of the combinations for ten, but not as an episode of working on fluency. It is interesting to note that, in my experience, this is very common (in France, at least): oral fluency of number facts is very frequently underestimated. More generally, orality (Goody 1977; Ong 2002), which has its own mode of knowing and organising facts, is not considered as really important. In this lesson, the very fact that the questions were presented as written sentences and organised in two columns, more (first column) or less (second column) organised by increasing one number, is somehow 'transparent' (Margolinas and Laparra 2011). There is frequently little awareness on the part of the teachers that oral mathematical facts and written ones are very different. For instance, in the Macao lesson, there was clearly a choice to be made: working on oral facts (and in this case giving questions orally and choosing questions with results not always ten) or working on organising facts about combinations for ten using writing. For some reason, oral calculation is not seriously considered, even if rapid oral calculation is still useful, whereas written calculation cannot compete with the use of a calculator: you will find more easily your phone in your pocket than paper and pencil.

The use of material (Stages 2–3) is also interesting because it might be found in different countries around the world, where base-ten material is generally used. The material used here has a property which is not always found: there is a ten-place grid composed of five and five places and you can take out or put elements in this grid. In some other classes, groups of ten cannot be decomposed: you thus have units and tens and if you have ten units you have to exchange those for a ten. There are discussions, around the world, on the differences between the two conceptions of place value: a ten is the first group in base-ten numeration (10^1) or a ten is a unit in itself. Choosing this material leads clearly to the first conception, but you can manipulate for at least two very different reasons. The first is purely material: if, for 24 + 9, a pupil shows two complete ten grids, one grid with four candies and one grid with nine candies, the teacher might say: 'you know that you are not allow to do that: you must complete the grid before taking another one'. In this case, pupils are only manipulating objects with a very loose relationship to base-ten operations, which is different if the teacher says: 'In order to organise the candies in base ten, you always have to make a group of ten when it's possible. Can you make another group of ten with your candies?'. The different solutions, which have been shown by the teacher in Stages 2 and 3, were clearly aimed at the second version, because they demonstrate different ways to regroup the candies in tens, using a schematisation. However, we do not know how to consider the relationship between boxes, candies and written numbers on the one hand and the role played by oral numeration on the other. This is particularly important when oral numeration is not congruent to written numeration (which is the case in the major European languages: you say twenty and not two-tens, where in most Asian languages oral numeration is regular). For instance, you can count ten, twenty, twenty-three (see figure in Stage 3) and write 23 as the cultural way to write twenty-three, or you can say two tens and directly write 2 in the left place which is the place value for tens and three units and write 3 in the right place (place value for units). With the same material, both decisions are possible, which are very different from a teaching point of view.

The selection of the introductory additions (40 + 3; 25 + 2; 25 + 20) highlights the teacher's choices and the mathematical knowledge of the team: in the first, a number with only tens and a number with only units are dealt with independently, in the second, you have to combine the units of the second number with those of the first and, in the last one, this is the same but with tens. Thus, the environment of the last problem, which is the core of the lesson (24 + 9), is not only material; it is also made of mathematical knowledge, which has been carefully introduced by the teacher. In focusing on more general considerations (existence of material, familiarity with the material, etc.), those calculations are components of the *milieu* (Brousseau 1997; Brousseau et al. 2014), which is never only material.

The teacher has determined in advance the possible procedures for 24 + 9. What is striking is that she has written everything in advance (Section). In this case, mathematical writing cannot emerge as a way to understand a solution, and there is no place for false solutions (see Ekdahl and Runesson 2015). Pupils might know the answer, either because other pupils have said it was 33 or because they have counted the candies one by one. Therefore, they might have the mathematical sentence right (24 + 9 = 33), but not the right base-ten properties. Wrong solutions might trigger the occasion to recall what base-ten is about: when you have ten, you regroup (which is true for units but will be true also for tens and so on). For example, with 4 and 9 you can make a ten, either with 4 + 6 (and leave 3) or with 9 + 1 (and leave 3), or you can know that 4 + 9 = 13, which is a ten and 3 units. Thus, it is an opportunity for the teacher to state the reasons for the three different solutions. This demonstrates the downside to having everything written in advance: the reasoning that underpins these solutions might remain unexplained.

The last remark relates to the conclusive part of the lesson. Task designers (Watson and Ohtani 2015) usually carefully describe the 'active' part of the task: the problem to solve and the environment of the problem. However, they usually avoid to enter into considerations about what you might tell pupils regarding what they have learned and what they have to memorise. If we use Brousseau's words, task designers are usually more concerned by the devolution process than by the institutionalisation process (Brousseau 1992; Margolinas 2005; Margolinas and Laparra 2008). The conclusive part in the Macao lesson shows clearly what the teacher expects of the pupils in the future: to learn the ten complements and to learn how to use them. The whole lesson appears, at this moment, as a whole, for students and for the observers.

Some Concluding Comments

Chapter and working group 4 have taken into consideration some important processes in teacher work. In an attempt to complement this work, I have focused on mathematical knowledge for teaching, in order to stress the need to consider our own conception of whole number arithmetic and the way it impacts our research and our analysis of teacher work.

If we take seriously the very interesting suggestion made in Sect. to-trans form a closed question into an open one (Sullivan and Lilburn 2004), we have thus to consider not only the shift in role it implies, but also the mathematical knowledge which might be learned by pupil difficulties and the mathematical knowledge necessary for the teacher. The intent of the Macao lesson, as clearly revealed in the concluding part, was to teach the use of the ten complements in order to give the result of any addition with carrying, which is useful either for mental or written calculations. The purpose of the study of the open problem proposed is completely different: it is true that it involves pupils doing additions and reflecting upon addition as an operation (and even as a function, since it can be modelled using the linear function $y = 33 - x$). The challenge for researchers might also be to find better problems with the *same* purpose, which is a very different question: that is, to focus also on 'daily routine' (see Brombacher 2015).

In general, I think that we often underestimate teacher knowledge required not only for selecting challenging and dense tasks but also, within a determined task, for responding to the diverse needs of individuals (Section) and to assess this need (see Gervasoni and Parish 2015). It is certainly not a little challenge for mathematics education research to describe the knowledge at stake, even within the field of WNA and even if we take a single lesson (see Lin 2015 for a development about the algorithm for multiplication). This book is certainly a very important step in this direction.

References

Ball, D. L., Thames, M. H., & Phelps, G. (2008). Content knowledge for teaching: What makes it special? *Journal of Teacher Education, 59*(5), 389–407.

Bikner-Ahsbahs, A., Prediger, S., & Halverscheid, S. (2014). Introduction to the theory of Interest-Dense Situations (IDS). In A. Bikner-Ahsbahs & S. Prediger (Eds.), *Networking of theories as a research practice in mathematics education* (pp. 97–113). Dordrecht/Heidelberg/New York/London: Springer.

Brousseau, G. (1992). Didactique: What can it do for the teacher. In R. Douady & A. Mercier (Eds.), *Research in Didactique of mathematics* (pp. 7–40). Grenoble: La Pensée sauvage.

Brousseau, G. (1997). *Theory of didactical situations in mathematics*. Dordrecht: Kluwer Academic Publishers.

Brousseau, G., Brousseau, N., & Warfield, V. (2014). *Teaching fractions through situations: A fundamental experiment*. Dordrecht/Heidelberg/New-York/London: Springer.

Goody, J. (1977). *The domestication of the savage mind*. Cambridge: University Press.

Goody, J. (1986). *The logic of writing and the organization of society*. Cambridge: Cambridge University Press.

Hill, H. C., Blunk, M. L., Charalambous, C. Y., Lewis, J. M., Phelps, G., Sleep, L., & Ball, D. L. (2008). Mathematical knowledge for teaching and the mathematical quality of instruction: An exploratory study. *Cognition and Instruction, 26*(4), 460–511.

Laparra, M., & Margolinas, C. (2009). Le schéma: un écrit de savoir? *Pratiques, 143–144*, 51–82. https://hal.archives-ouvertes.fr/hal-00722211.

Margolinas, C. (2005). La dévolution et le travail du professeur. In P. Clanché, M.-H. Salin, & B. Sarrazy (Eds.), *Autour de la théorie des situations* (pp. 329–333). Grenoble: La Pensée sauvage.

Margolinas, C., & Laparra, M. (2008). *Quand la dévolution prend le pas sur l'institutionnalisation. Des effets de la transparence des objets de savoir.* Les didactiques et leur rapport à l'enseignement et à la formation. http://hal.archives-ouvertes.fr/hal-00779656

Margolinas, C., & Laparra, M. (2011). Des savoirs transparents dans le travail des professeurs à l'école primaire. In J.-Y. Rochex & J. Crinon (Eds.), *La construction des inégalités scolaires* (pp. 19–32). Rennes: Presses universitaires de Rennes.

Ong, W. J. (2002). *Orality and literacy*. London/New York: Routledge.

Shulman, L. S. (1986). Those who understand. Knowledge growth in teaching. *Educational Researcher, 15*(2), 4–14.

Sullivan, P., & Lilburn, P. (2004). *Open-ended maths activities: Using "good" questions to enhance learning in mathematics*. Melbourne: Oxford University Press.

Vergnaud, G. (1983). Cognitive and developmental psychology and research in mathematics education: Some theoretical and methodological issues. *For the Learning of Mathematics, 3*(2), 31–41.

Watson, A., & Ohtani, M. (2015). *Task design in mathematics education*. Heidelberg: Springer.

Cited papers from Sun, X., Kaur, B., & Novotna, J. (Eds.). (2015). Conference proceedings of the ICMI study 23: Primary mathematics study on whole numbers. Retrieved February 10, 2016, from www.umac.mo/fed/ICMI23/doc/Proceedings_ICMI_STUDY_23_final.pdf

Alafaleq, M., Mailizar, L., Wang, Y., & Fan, L. (2015). How equality and inequality of whole numbers are introduced in China, Indonesia and Saudi Arabia primary mathematics textbooks (pp. 392–398).

Askew, M. (2015). Seeing through place value: An example of connectionist teaching (pp. 399–406).

Barry, A., Novotná, J., & Sarrazy, B. (2015). Experience and didactical knowledge: The case of

didactical variability in solving problems (pp. 407–414).
Brombacher, A. (2015). National intervention research activity for early grade mathematics in Jordan (pp. 415–422).
Cao, Y., Li, X., & Zuo, H. (2015). Characteristics of multiplication teaching of whole numbers in China: The application of the nine times table (pp. 423–430).
Ekdahl, A.-L., & Runesson, U. (2015). Teachers' responses to incorrect answers on missing number problems in South Africa (pp. 431–439).
Gervasoni, A., & Parish, L. (2015). Insights and implications about the whole number knowledge of grade 1 to grade 4 children (pp. 440–447).
Kaur, B. (2015). The model method: A tool for representing and visualising relationships (pp. 448–455).
Lin, P.-J. (2015). Teaching the structure of standard algorithm of multiplication with 2-digit multipliers via conjecturing (pp. 456–463).
Pearn, C. (2015). Same year, same school, same curriculum: Different mathematics results (pp. 464–471).
Sensevy, G., Quilio, S., & Mercier, A. (2015). Arithmetic and comprehension at primary school (pp. 472–479).
Wong, I.N., Jiang, C., Cheung, K-C., & Sun, X.H. (2015). Primary mathematics education in Macao: Fifteen years of experiences after 1999 handover from Portugal to mainland China (pp. 480–487).
Zhang, Q.-P., Cheung, K.-C., & Cheung, K.-F. (2015). An analysis of two-digit numbers subtraction in Hong Kong primary mathematics textbooks (pp. 488–495).
Zhao X., van den Heuvel-Panhuizen, M., & Veldhuis, M. (2015). Classroom assessment techniques to assess Chinese students' sense of division (pp. 496–503).

3

Social and Cultural Contexts in the Teaching and Learning of Whole Number Arithmetic

Maria G. Bartolini Bussi and Xu Hua Sun

Introduction

At the first meeting of the ICME 23 IPC group, held in Berlin in January 2014, there was unanimous agreement on the relevance of cultural diversity in ICMI Studies. The Discussion Document for the study noted that:

> It was decided that cultural diversity and how this diversity impinges on the early introduction of whole numbers would be one major focus. The Study will seek contributions from authors from as many countries as possible, especially those in which cultural characteristics are less known and yet they influence what is taught and learned. In order to foster the understanding of the different contexts where authors have developed their studies, each applicant for the Conference will be required to prepare background information (on a specific form) about this context. (see this volume, Appendix 2, "Introduction and Rationale for ICMI Study 23")

This statement was based, on the one hand, on the awareness of the increasing participation of scholars from developing countries in international conferences and of the number of submissions to international journals of manuscripts from all over the world and, on the other hand, on the ICMI aim to improve the quality of mathematics teaching and learning worldwide. Most IPC members, including the first author of this chapter, have experience in reviewing papers for international

---

The second author was supported by the Research Committee, University of Macau, Macao, China (MYRG2015-00203-FED). The opinions expressed in the article are those of the authors.

M.G. Bartolini Bussi (✉)
Department of Education and Humanities, University of Modena and Reggio Emilia, Modena, Italy
e-mail: mariagiuseppina.bartolini@unimore.it

X.H. Sun (✉)
Faculty of Education, University of Macau, Macao, China
e-mail: xhsun@umac.mo

conferences and journals: in many papers, there is an implicit belief that the readership knows enough about the context in which a study has been carried out (especially if it concerns European or North American countries) and that the transposition of findings from one country to another is possible and natural, if the theoretical framework and methodology are sound enough.

However, as early as the ICMI Study 8 (1992–1998) on *Mathematics Education as a Research Domain: A Search for Identity* (Sierpinska and Kilpatrick 1998), working group 4, led by Susan Pirie, Tommy Dreyfus and Jerry Becker, was raising questions about the issue of results and their validity. An interesting question was raised:

> To what extent can research results from one environment or culture (e. g. Japan) be linked to those from another culture (e.g. the USA) and to what extent are results culture specific? (p. 27)

Although the issue was raised about 20 years ago, at the ICMI Study Conference in Washington, DC, in 1994, the acknowledgement of the issue of cultural context in major international journals and conferences remains uncommon (Bartolini Bussi and Martignone 2013). In support of this claim, it is enough to quote some excerpts from the information given to prospective authors of empirical studies by one of the major international journals, i.e. the *Journal for Research in Mathematics Education* (NCTM n.d.).

> The Journal for Research in Mathematics Education seeks high quality manuscripts that contribute knowledge to the field of mathematics education. For an author's work to be publishable, it needs to exhibit qualities that characterize well-conceived and well-reported research studies. The following information illustrates characteristics of strong manuscripts that have been submitted to JRME.

The following items are then elaborated:

> appropriate purpose and rationale
> clear research questions
> an informative literature review
> a coherent theoretical framework
> clearly described research methods
> sound research design and methods
> claims about results and implications that are supported by data
> contribution to the field of mathematics education
> clearly explained and appropriately used terms
> high quality writing
> mathematical accuracy.

An abridged version of the same document appears in the PME 39 (2015) guidelines for research reports of empirical studies.

> Observational, ethnographic, experimental, quasi-experimental, and case studies are all suitable.
>
> Reports of empirical studies should contain, at minimum, the following:
>
> - a statement regarding the focus of the submitted paper;
> - the study's theoretical framework;
> - references to the related literature;

- an indication of and justification for the study's methodology; and
- a sample of the data and the results (additional data can be presented at the conference but some data ought to accompany the proposal)

In both cases, no reference to the social and cultural context is explicitly mentioned. Hence, the limited space allowed either for manuscripts or (even more limited) for research reports is likely to inhibit the author's intention of framing the empirical study within its context. Moreover, it implicitly conveys the idea that every relevant scientific communication must follow the above structure, where there is no reference to the social and cultural context. It seems a limiting rather than a proactive statement.

For example, open-class activities/lesson studies that view the classroom as an open or public space, which has been a major influence in the professional development of teachers in China and Japan for many years, may be contrasted with the view of the private and autonomous classroom that has been described as common in the Western tradition (see, for instance, Sztein et al. 2010).

Instead, the evidence of some ongoing changes may be found in the guidelines for reviewers in two subsequent Conferences of the European Society for Research in Mathematics Education. The CERME 8 (2013) guidelines read:

Reports of Studies (Empirical or Developmental)
Surveys, observational, ethnographic, experimental or quasi-experimental studies, case studies are all suitable. Papers should contain at least the following:
a statement about the focus of the paper;
an indication of the theoretical framework of the study reported, including references to the related literature;
an indication of and justification for the methodology used (including problem, goals and/or research questions; criteria for the selection of participants or sampling; data collection instruments and procedures);
results;
final remarks or conclusions

The CERME 9 (2015) guidelines added a new indicator:

an indication on the scientific and cultural context in which this study is embedded (explaining crucial assumptions and the possible contingency of the relevance of the study for a specific cultural context)

What happened between the CERME 8 and CERME 9 guidelines? Both co-chairs of this ICMI23 Study were present at CERME 8, and the first author of this chapter was invited by the organising committee to introduce a forum discussion about the neglected importance of the social and cultural context. The CERME board was favourably impressed and accepted the challenge to adapt the widespread tradition mentioned above. CERME 9's scientific committee not only introduced a small change in the instruction (for authors and reviewers), but also decided to host a panel chaired by Barbara Jaworski on *Cultural Contexts for European Research and Design Practices in Mathematics Education* (Jaworski et al. 2015). The panel was held very successfully in Prague in February 2015. This panel represents a milestone in making explicit international awareness of the importance of the social and cultural context in the teaching and learning of mathematics.

Another milestone is represented by the explicit request for contextual detail in the ICMI Study 23. The ICMI Study IPC unanimously agreed to highlight the role of the social and cultural context and to design a specific form (see below) for collecting relevant information about this context, in order to leave the limited space of the paper (eight pages) for the scientific report according to the usual formats. The aim was twofold: not only to collect relevant information for understanding the different contexts (as explicitly written in the Discussion Document), but also to foster authors' awareness about the relevance of their own cultural contexts.

The Context Form: Design

The form designed by the IPC tried to address some very basic issues about the situation in the country where either the empirical research study or the theoretical reflections were carried out. The IPC was aware that a complete answer to all the questions would have been very demanding, and akin to a study itself, unless the authors knew some already existing documents at the national level (e.g. ICMI 2011).

The form designed by the IPC follows (Table).

Some applicants expressed surprise at this unexpected additional task, asking the reasons for completing such a form with information that should already be known by all mathematics educators: this was further evidence, if any were needed, that the awareness of the relevance of the social and cultural context and the need to offer information about these are far from being shared in the field.

The following story of the Macao Conference, with the visit to Chinese schools and with the discovery of the differences between Western and Eastern mathematical traditions, provided further evidence that this awareness is really needed and useful in order to understand and to start a fruitful dialogue between different cultural contexts. This aspect will be further elaborated in this chapter and in the whole volume.

The Context Form: Data

Sixty-six context forms were collected, concerning 29 countries (counting separately China and SARs Hong Kong and Macao). The distribution by country is detailed in Table.

Three submitted papers concerned cross-cultural studies (Cyprus – Netherlands, Germany – Australia, England – Sweden, where the context forms for both countries were filled (and, hence, were counted twice in the above table). In one case, the submitted paper concerned a cross-cultural study where all the Francophone countries were analysed for a study commissioned by the World Bank: a single context

Table The context form

Please fill in as many of the following as completely as possible so that we understand the context of your paper	
General (objective data)	Give a rough idea of the numbers of:
	Students up to the age of 11 years
	Teachers for pre-primary and primary education
	A short description of the National Education System (please match grades with pupils' age)
	If relevant, please explain whether the system is inherited from colonial period or is related to local traditions
	Give information about what you consider an important feature in your country (e.g. the pillar of the network of monastic schools in Burma)
Inclusiveness	Is the system totally inclusive?
	Are there special schools/classrooms for
	sensually impaired students (blind, deaf)?
	Are there special schools for students with disabilities?
National language(s)	List the national language(s) of the country
	List the local languages (minorities)
	Is pre-primary/primary school carried out in the local languages?
Migrant/refugee/marginalised students	Is there a significant minority of migrating students (coming from other countries), of refugee students and of marginalised students?
	Are there specific rules for schools which take care of these students? Is there some help from the national/local government?
Pre-primary general	Is pre-primary education extended to the whole country?
	Which percentage of students are expected to enrol in pre-primary education?
Pre-primary textbooks	Do they exist?
	In what language (in the case of more local languages)?
	Is there only one national textbook? Or a limited number?
	Is there only one teachers' guide? Or a limited number?
Primary general	Is primary education extended to the whole country?
	Which percentage of students are expected to enrol in primary education?

(continued)

Table (continued)

Please fill in as many of the following as completely as possible so that we understand the context of your paper	
Primary textbooks	Do they exist? In what language (in the case of more local languages)?
	Is there only one national textbook? Or a limited number?
	Is there only one teachers' guide? Or a limited number?
Primary teachers' qualification	Generalists or specialists?
Assessment	Is there a national system of assessment?
	At what ages/grades are students assessed in mathematics (focus on both pre-school and primary school)?
Standards	Is there a governmental/national document for standards?
Teacher education and development	What are the national rules?
	Is there some shared practice you consider relevant (e.g. Lesson Study in Japan, guānmó kè in China)?
	Do you have forms of distance learning for teacher development?
Teacher education and development: pre-primary	How is organised pre-primary teacher education?
	Please distinguish (if relevant) the governmental rules and what happens in practice?
Teacher education and development – primary	How is organised primary teacher education?
	Please distinguish (if relevant) the governmental rules and what happens in practice?
Contents (limiting the focus to whole numbers)	Local languages
	Place value: do you have tradition of system of representation in base not ten?
	Problems: which kind of problems are typical of school practice? (e.g. China: problems with variation)
	Problems: which kind of problem-solving strategies? (e.g. Singapore: model method)?
Any other information related to the context of your paper	

form was filled for all the Francophone countries, with reference to the colonial influence of the French system.

The applicants were encouraged to fill the form as completely as possible, drawing on their own knowledge. Hence, rather than on objective data, in most cases, the information drew on applicants' knowledge and perceptions of their national contexts. Moreover, the sample was a convenience sample involving the selection of the most accessible subjects (Marshall 1996), limited to the applicants in the study,

Table The countries

Country	Number of forms	Country	Number of forms
Algeria	1	Australia	5
Belgium	1	Brazil	1
Canada	4	China	5
China HK SAR	2	China Macao SAR	2
Cyprus	1	Czech Republic	1
Denmark	1	Dominican Rep.	1
England	1	France	4
Germany	4	Israel	3
Italy	4	Jordan	1
Netherlands	2	New Zealand	2
Serbia	1	Singapore	1
South Africa	3	Sweden	2
Switzerland	1	Taiwan	4
Thailand	1	USA	5
Vietnam	1	All Francophone countries	1

hence excluding or limiting the contribution of some major areas (e.g. India, Russia, Latin America, much of Africa, Southeast Asia).

In the following, we briefly outline the main outcomes of an early analysis of the collected data.

The General Structure of Education Systems for Early Years Mathematics

The data reported by the applicants have been matched with Education Database (n.d.). Although in some countries (e.g. Australia, Brazil, Canada, Cyprus, Germany, Switzerland, USA) there are differences between the different states/provinces/regions/territories, the models may be summarised as follows. While primary school or elementary school is the accepted wording, in some cases pre-primary school is named in different ways. Usually pre-primary is not mandatory but attended by many students (in some cases up to 95%) at least in the last year before entering primary school.

In some cases primary school is split into different steps including also what is elsewhere called middle school. Although our convenience sample is limited to 29 countries, there is a large variety, concerning both the duration and the entry age. In Europe too different models exist. This institutional diversity has implications for this study: for instance, when the entry age is postponed, it is likely that WNA is approached at pre-school level, and when the duration is extended (up to sixth grade, as in many Eastern countries), it is likely that pre-algebraic thinking is fostered before high school level. As these institutional differences cannot be cancelled, in this study we chose to focus on the contents rather than on the grades or the students' age (Table).

The influence of colonial heritage is reported in some countries: Algeria, Australia, China HK, China Macao, New Zealand, Taiwan (from Japan and China). This influence in some cases emerges also in the choice of school language different from family language. This issue was reported and discussed also in the Conference (this volume, Chap.).

Inclusiveness in Education

The focus on inclusive education has a long story in UNESCO's documents (e.g. UNESCO 2009a; see the historical summary on p. 9) and dates back to the Universal Declaration of Human Rights (1948), but it is still actual (see, for instance, the plenary speech by Bill Barton in ICME 13[1]). Inclusive education was taken into account in the *Millennium Developmental Goals* criteria (UNESCO 2010), where *Universal Primary Education* (UPE) is mentioned (Millennium Developmental Goal 2). It is considered also in the most recent document (UNESCO 2017) on *Education for Sustainable Development Goals,* where it is included in the learning objective 4: 'Quality Education I Ensure inclusive and equitable quality education and promote lifelong learning opportunities for all' (UNESCO 2017, p. 18 ff.).

UNESCO (2009a) states that inclusion addresses not only students with special needs[2] (e.g. disable students), but also those from diverse backgrounds (cognitive, ethnic and socio-economic). Hence, this issue is related to some other questions posed in the context form (i.e. national languages and school languages, provisions for migrant, refugee and marginalised students).

This broad approach was assumed by the IPC of the study which included in the context form three different items (see above) concerning students with special needs, students with school language different from family language and students with diverse background.

The presence of different national languages has been reported by applicants, mentioning also local languages. There are countries where home languages are different from national language (or languages), for instance, in Algeria, Arabic and Berber–Tamazight; in Australia, Australian English and aboriginal languages; in Belgium, Dutch, French and German; in Canada, English and French; in the Chinese

[1] https://lecture2go.uni-hamburg.de/l2go/-/get/v/19757)

[2] We are aware that there is a growing trend towards abandoning the wording 'special needs' and using 'special rights' or 'educational rights'. For instance, Runswick-Cole and Hodge (2009) have argued for abandoning the language of 'special education needs' in the UK, based on the claim that it has led to exclusionary practices, mentioning an Italian rights-based approach developed in Reggio Emilia that refers to 'children with special rights' drawing on the United Nations Convention of the rights of the child (UNICEF 1989). Yet we decided to maintain the most common wording 'special needs' as it is shared in literature and better known by mathematics educators.

Table Structure of primary school

4-6 age: 6–10 (12)	5 age: 6–11	5–6 age: from 6	6 age: from 5	6 age: 6–12	6 age: 7–13	5–6–7 age: from 6	7 age: 7–14	8 age: 4–12	8 age: 7–15	9 age: 6–15	9 age: 7–16	10 age: from 6–16
Germany (depending on region)	*Algeria* (under colonial influence)	*Cyprus* (Greek, 6; Turkish, 5)	*UK* (complex system)	*Australia*	*Taiwan*	*Canada* (depending on territories)	*South Africa*	*Netherlands*	*Serbia*	*Brazil* (5+4)	*Denmark*	*Jordan*
	France			*Belgium*	*Singapore*		*Switzerland* (complex system)			*Czech Rep.*	*Sweden*	
	Italy			*China*						*Dom. Rep.*		
	Vietnam			*Hong Kong* (modelled after UK system)						*New Zealand*		
	USA (complex system)			*Macao* (under different influences)								
				Israel (complex system based on religions)								
				Thailand								

area, Mandarin, Cantonese and minorities languages; in Cyprus, Greek and Turkish; in Israel, Hebrew and Arabic; in New Zealand, English, Te Reo Maori and New Zealand Sign Language; in Serbia, Serbian, Hungarian and Romanian; in Singapore, English, Malay, Tamil and Mandarin; in South Africa, Afrikaans, English, Zulu, Xhosa, Swati, Tswana, Southern Sotho, Northern Sotho, Tsonga, Venda and Ndebele; in Sweden, Swedish, Finnish, Meankeli, Samic, and so on; in Switzerland, French, German, Italian and Romansh; in Thailand, Thai and Esann; and in the USA, English and Spanish. In most cases, the language of teaching (or school language) is different from home languages with the well-known critical consequences (Barwell et al. 2016). There are countries with acknowledged minorities (e.g. Czech Republic, France, Italy, New Zealand) where teaching in the minority language is encouraged with special funds and programmes. For instance, in New Zealand, Maori schools are very well developed and address about 15–20% students.[3]

The issue of migrant, marginalised and refugee students is mentioned by some applicants, although only in a few countries (e.g. Australia, Belgium, Cyprus, Germany, Jordan, Netherlands, New Zealand) official governmental support is mentioned. In other cases (e.g. France, Italy, UK), municipal support is mentioned together with the involvement of volunteers and charities.

According to the data reported in UNESCO (2017) and confirmed by some participants, there has been significant progress towards ensuring UPE in terms of access, and the conversation has now shifted from aiming for access to goals for quality UPE.

The question of students with disabilities or special needs seems to be ill-posed or, maybe, ill-interpreted by the applicants to the Conference. In many cases, applicants answered YES (i.e. the system is inclusive) probably meaning that all the students are allowed to go to primary school, but in many cases (at least 12 out of 29), special schools for disabled students were mentioned as the only provision.

According to UNESCO (2009a):

> In most countries, both developed and developing, the steps towards achieving the right to education for students with disabilities have followed a common pattern, with some local variations. Progress has tended to follow the pattern of steps outlined below:
>
> **Exclusion** from school, based on negative attitudes and a denial of rights, justified by the belief that students with disabilities cannot learn or benefit from education
>
> **Segregation**, reflecting the emphasis on 'difference', combined with a charity-based approach, where separate education centres and schools were and are still provided by local, regional and international charitable NGOs and, more recently, by development-focused NGOs
>
> **Integration**, reflecting some degree of acceptance for some disabled students, depending on their degree of disability, allowing them to attend local regular national schools, as long as they can fit in to the school and the school does not have to make significant adjustments for them
>
> **Inclusion** in education, acknowledging the fact that all students, including those with disabilities, have the right to education, that all schools have the responsibility to teach every child and that it is the responsibility of the school to make the adjustments that may be necessary to make sure that all students can learn (p. 51)

[3] www.education.govt.nz/ministry-of-education/our-role-and-our-people/education-in-nz/

Following this definition, if the system of special schools is widespread in a country, it is a *segregation* model and not an *inclusive* model.

To sum up the data, most countries are reported to have special schools only; some countries (e.g. Australia, Brazil, Canada, Czech Republic, Denmark, France, Germany, Israel, Netherlands, New Zealand) have started a process of *integration*, and some countries have stated by law complete *inclusion* in mainstream classes with support teachers. A relevant case, worthwhile to be mentioned, is Italy. D'Alessio (2011) reconstructs the historical and legislative backgrounds of the integration policy in Italy, mentioning the promulgation of the Italian Constitution (Senato della Repubblica 1947), where the spirit and ethos for integration were already encapsulated.

> Since the Fascist dictatorship had denied individual freedom, one of the first targets of the democratic Constitution was to put the dignity of the person and the rights of minorities at the centre of the constitutional charter. (D'Alessio 2011, p. 6)

In the following years, legislation went in this direction. According to Ferri (2008), for Italian teachers, inclusion was considered 'a moral issue which is more important than a legal mandate' (p. 47). A discussion about the possible distance between laws and implementation is made by Booth and Ainscow (2011) which have designed a tool to support and assist with the process of developing inclusive education.

At the international level, inclusive education is considered to be:

> a key vehicle through which the right to an equal education opportunity for all can be ensured. For this to become a reality it is necessary to provide a system in which all persons, including persons with disabilities, can access education at all levels on an equal basis with others in the communities in which they live. They should not be excluded on the basis of any disability and should get the support they require. (EASPD 2012, p. 6)

It must be said, however, that the issue of *exclusion-segregation-integration-inclusion* is far from being agreed upon at the international level. It is not only a matter of clear definitions; it is rather a matter of ethical consensus. A recent paper by Reindal (2016) summarises different positions, reconstructing the history of the inclusion debate from the 'World Conference on Special Needs Education' in Salamanca in 1994. Reindal claims:

> Inclusive education as presented in documents from UNESCO was indefinite from the start in relation to both the target group and those whose responsibility was to implement inclusive education for that group. Reviews of the research also support several interpretations of responsibility and elucidations of inclusive education. In a recent study based on prior reviews and a recent search of databases covering the period 2004–2012, Goransson and Nilholm (2014a) found four different interpretations of inclusion which gave rise to four qualitatively different categories of definitions. These definitions were related hierarchically to each other employing stricter criteria concerning what counts as inclusive education as one goes from A to D:

(A) Placement definition – inclusion as the placement of pupils with disabilities in mainstream classrooms.

(B) Specified individualised definition – inclusion as meeting the social/academic needs of pupils with disabilities.

(C) General individualised definition – inclusion as meeting the social/academic needs of all pupils
(D) Community definition – inclusion as creation of communities with specific characteristics.

Reindal (2016) suggests to tackle this issue from the perspective of a *capability approach* (Walker and Unterhalter 2007), as that:

> has the potential to emphasise the ethical aspects of inclusion because it builds on an understanding of difference as a specific variable of human diversity, and because it understands human dignity as the development of capabilities. The capability approach defends an understanding of difference as a specific variable of human diversity with an objective reality. [...] If the central purpose of special education and inclusion is to treat all students as the same while at the same time aiming to treat them differently then one must deal with the problem of difference in a way that comes to grips with the attendant challenges – as well as those faced particularly by developing countries. (Reindal 2016, p. 6)

The issue of the diversity of school language and family language was discussed during the Conference and finds place in this Volume The issue of students with special needs was discussed during the Conference and reported in some chapters.

The capability approach, one that is very interesting, was not picked up in the Study and may suggest future developments in mathematics education.

Textbooks

Most applicants reported that no textbook for pre-primary exists: rather, some available collections of learning resources, working sheets and teachers' guides are mentioned.

As far as primary school is concerned, textbooks exist everywhere, although in some cases (e.g. Australia) the adoption of a textbook is not mandatory. In most countries there is a free-market system with no official overlooking agency. In some countries only one or a limited number of approved textbooks is available (e.g. Chinese area, Algeria, Germany, Singapore and Thailand). In Vietnam, textbooks are written by specialists of the Ministry of Education. In South Africa, there are textbooks in all the national languages, though harder to access in the smaller language groups. In China too, there are textbooks in all the minority languages.

In this Study, the issue of textbooks was just skimmed in the Chapters but would deserve a study in its own.

National Curriculum Standards and Assessment

In nearly all the countries of this convenience sample, there are national standards. They do not exist in Algeria and are identified with the sole national textbook in Jordan. An interesting case is represented by the USA. There is no national curriculum in the USA, but National Council of Teachers of Mathematics (NCTM) Standards are widespread. Yet, locally, states, school districts and national associations recommend some curriculum standards be used to guide school instruction. The Common Core Standard initiative[4] is in progress. In nearly all the countries, there is a national system of assessment (in progress in New Zealand) consistent with curriculum standards. However, the grades in which assessment takes place are not the same. Usually they are every second or third year and depend also on the structure of the education system.

Teachers' Qualification and Teacher Education and Development

All the applicants reported a *generalist* trend for primary teacher education in their countries, with some limited exceptions: Germany, with the encouragement to get further qualification in German or Mathematics, and Italy, with testing of specialist mathematics teachers in some schools, according to autonomous choices of the school council. Only in Denmark and in the Chinese area a trend towards specialist mathematics teachers is reported. In China, this choice is common in big cities but not in rural areas. Where there are specialist mathematics teachers, it is common to form a Mathematics Teaching Research Group in the school, for in-service development according to the model of 'open classes', called in Chinese *guānmó kè* (观摩课), which means 'to observe for imitating a lesson' and has some similarities to the Japanese lesson study (Sun et al. 2015).

In our convenience sample, pre-service teacher education at universities (or, in some cases, teachers' colleges) seems well established. In most cases for pre-primary and primary school, the length of the programme (bachelor) is the same:

3-year bachelor in Belgium and New Zealand
4-year bachelor in Australia, Canada, Chinese area, South Africa, Switzerland
5-year master in Italy
6-year master in Thailand
In other countries, the length is different for pre-primary and primary school teachers:
2–3 years in Singapore (at the National Institute of Education)
3–4 years in Denmark, Serbia
3 1/2–4 years in Sweden
In some cases the length of the bachelor degree is not reported.
In Germany no university programme for pre-primary teachers is reported, while the programme for primary teachers is at masters level (5 years).
In France there is no information provided for pre-primary, while the programme for primary teachers is at masters level (5 years).

[4] www.corestandards.org

In the USA the rules are different in different states. For primary teachers, bachelor degrees are mentioned.
In the UK, different agencies are involved with programmes of different lengths (e.g. school-led teacher training, university programmes).

Practicum (internship) is mentioned with very different organisations and durations, for instance, a 600 hours practicum is required in Italy alongside the 5-year master programme, while in Israel, the practicum is at the end of the university programme. A year of practicum is required in Macao alongside the 4-year bachelor degree. In Australia 80 hours of mandatory practicum is prescribed for bachelor programmes.

Distance learning is mentioned especially for in-service teacher development. However, for in-service development, different models are described from inspector-led programmes, to mandatory programmes (60 hours per year in Israel, 5 days per year in Switzerland). In Australia, there is a programme for accreditation according to well-described professional standards. In some cases in-service development is appointed to municipalities (Sweden). In many cases in-service development is reported as not effective. It seems that, in general, there are not well-organised models. A relevant exception is the model of 'open classes' in China.

For teacher education and development, besides the panel on teacher education (this volume, Chap.), it is worthwhile to mention the ICMI Study 15 (Even and Ball 2009).

Conclusion

In this chapter we have briefly explored mainly structural features of the instruction systems including inclusiveness, curricula, standards and assessment and teacher education and development. We had just a glance to the many different choices existing in the countries of our convenience sample.

Our limited analysis shows that even in Europe, a small continent, many different organisations of the education systems exist. Education systems are cultural artefacts that could be studied, on the one hand, as products of socio-cultural contexts and, on the other hand, as sources of information about the society that constructed or adopted them. We do hope that, in the future, a sensitive attitude for cultural contexts will become more and more shared, in international journals and conferences. As stated by Bartolini Bussi and Martignone (2013):

> The question of cultural background applies to every study in mathematics education [...] It is necessary to explain in more depth how the research design and implementation is related to the cultural background: the results and success (if any) of the project may depend on implicit values which are not likely to be found in other contexts. (p. 2)

In this spirit, the investigation of the different social and cultural contexts continued throughout the whole study and is mirrored in the Study Volume.

References

Bartolini Bussi, M. G., & Martignone, F. (2013). Cultural issues in the communication of research on mathematics education. *For the Learning of Mathematics, 33*(1), 32–38.

Barwell, R., Clarkson, P., Halai, A., Kazima, M., Moschkovich, J., Planas, N., Phakeng, M., Valero, P., & Villavicencio Ubillús, M. (2016). *Mathematics education and language diversity: The ICMI study 21st*. Cham: Springer.

Booth, T., & Ainscow, M. (2011). *Index for Inclusion: Developing learning and participation in schools* (3rd revised ed.). Canterbury: Centre for Studies on Inclusive Education.

CERME 8. (2013). http://cerme8.metu.edu.tr/#. Retrieved January 24, 2016.

CERME 9. (2015). http://www.cerme9.org/guidelines/. Retrieved January 24, 2016.

D'Alessio, S. (2011). *Inclusive education in Italy. A critical analysis of the policy of "integrazione scolastica"*. Rotterdam: Sense Publishers.

EASPD. (2012). *Analysis of the use and value of the Index for Inclusion (Booth and Ainscow 2011) and other instruments to assess and develop inclusive education practice in P2i partner countries*. Brussels/Tilburg, Fontys OSO. http://www.icevi-europe.org/enletter/issue51-09EASPD3.pdf. Retrieved December 27, 2016.

Education database. (n.d.). http://www.classbase.com/Countries. Retrieved January 24, 2016.

Even, R., & Ball, D. L. (2009). *The professional education and development of teachers of mathematics: The 15th ICMI study*. New York: Springer.

Ferri, B. (2008). Inclusion in Italy: What happens when everyone belongs. In S. L. Gabel & S. Danforth (Eds.), *Disability and the politics of education: An international reader* (pp. 41–52). New York: Peter Lang Publishers.

Göransson, K., & Nilholm, C. (2014). Conceptual diversities and empirical shortcomings: A critical analysis of research on inclusive education. *European Journal of Special Needs Education, 29*(3), 265–280. https://doi.org/10.1080/08856257.2014.933545.

ICMI (2011). *The ICMI database project*. Http://Www.Mathunion.Org/Icmi/Activities/Database-Project/Introduction/. Retrieved January 24, 2016.

Jaworski, B., Bartolini Bussi, M. G., Prediger, S., & Nowinska, E. (2015). Cultural contexts for European research and design practices in mathematics education. In K. Kreiner & N. Vondrova (Eds.), *CERME9. Proceedings of the Ninth Congress of the European Society for Research in Mathematics Education* (pp. 7–33). Prague: Charles University in Prague, Faculty of Education and ERME.

Jones, K., Bokhove, C., Howson, G., & Fan, L. (2014). *Proceedings of the International Conference on Mathematics Textbook Research and Development*. Southampton: University of Southampton. Retrieved June 2, 2016 from http://eprints.soton.ac.uk/374809/

Marshall M. N. (1996) Sampling for qualitative research. *Family Practice, 13*(6), 522–525.

NCTM. (n.d.). *Characteristic of a high quality JRME manuscript*. http://www.nctm.org/publications/write-review-referee/journals/Characteristics-of-a-High-Quality-JRME-Manuscript/. Retrieved January 24, 2016.

PME 39. (2015). *Submissions: Research reports*. http://www.pme39.com/submissions/. Retrieved January 24, 2016.

Reindal, S. M. (2016). Discussing inclusive education: An inquiry into different interpretations and a search for ethical aspects of inclusion using the capabilities approach. *European Journal of Special Needs Education, 31*(1), 1–12.

Runswick-Cole, K., & Hodge, N. (2009). Needs or rights? A challenge to the discourse of special education. *British Journal of Special Education, 36*(4), 198–203.

Senato della Repubblica. (1947). *Constitution of the Italian Republic*. https://www.senato.it/documenti/repository/istituzione/costituzione_inglese.pdf. Retrieved 10 December 2016.

Sierpinska, A., & Kilpatrick, J. (1998). *Mathematics education as a research domain: A search for identity, An ICMI study*. Dordrecht: Springer.

Sun, X.H., Teo, T., & Chan, T.C. (2015). Application of the open-class approach to pre-service teacher training in Macao: A qualitative assessment. *Research Papers in Education, 30*(5), 567–584.

Sztein, E., Olson, S., & Ferreras, A. (Eds.). (2010). *The teacher development continuum in the United States and China: Summary of a workshop*. Washington, DC: The National Academies Press.

UNESCO. (2009a). *Towards inclusive education for students with disabilities: A guideline.* Bangkok: UNESCO Asia and Pacific Regional Bureau for Education. http://unesdoc.unesco.org/images/0019/001924/192480e.pdf. Retrieved October 25, 2017.

UNESCO. (2009b). *Policy guidelines on inclusion in education*. Paris: United Nations Educational, Scientific and Cultural Organization. unesdoc.unesco.org/images/0017/001778/177849e.pdf. Retrieved April 24, 2017.

UNESCO. (2010). *Resolution adopted by the General Assembly. 65/1. Keeping the promise: United to achieve the Millennium Developmental Goals*. www.un.org/en/mdg/summit2010/pdf/outcome_documentN1051260.pdf. Retrieved April 24, 2017.

UNESCO. (2017). *Education for sustainable development goals*. Learning objectives. http://unesdoc.unesco.org/images/0024/002474/247444e.pdf. Retrieved April 24, 2017.

UNICEF. (1989). *Convention on the rights of the child.* www.unicef-irc.org/portfolios/crc.html. Retrieved April 25, 2017.

Walker, M., & Unterhalter, E. (2007). The capability approach: Its potential for work in education. In M. Walker & E. Unterhalter (Eds.), *Amartya Sen's capability approach and social justice in education* (pp. 1–18). London/New York: Palgrave Macmillan.

4

On Number Language

David Pimm

Introduction

I start with a policy statement, pretty unrelated to the previous chapter. I am always a little taken aback to see numbers or other mathematical symbols (e.g. '7' rather than 'seven', '+' rather than 'plus') presented inside classroom transcripts, which supposedly provide a written account of what was *said*. Everything that is said is said by someone in some natural language (or natural language mix – cf. code-switching, e.g. Setati 1998 – such as where a somewhat bilingual speaker may know how to say the higher number words in one language only). Non-verbal numerals (of whatever sort) are *not* part of any natural language,[1] so they require ways to be read aloud *into* such a language. Because of this, I believe it is important to be very, very precise about marking such distinctions. In Pimm (1987), for instance, I distinguished between what I termed a 'spelling' reading and an 'interpretative' reading of written mathematics: for example, is the Biblical 'number of the beast' (666) to be said as 'six six six' or is it 'six hundred and sixty-six' (in British English) or 'six hundred sixty-six' (in the North American version)? What it is not, however, is 'six hundreds (and) sixty-six', something I will come back to later on.[2]

[1] Chrisomalis asserts, 'Over 100 structurally different numerical notation systems are known to have been used between 3500 BCE and the present day [...] Unlike number words, they represent numbers translinguistically, and do not follow the language or lexicon of any specific language. Unlike tallies, they represent completed enumerations, and unlike computational technologies, they create permanent records of numerals' (Chrisomalis 2009, pp. 506–7).

[2] Note this is not true when saying decimals in English: '666.66' can be read aloud as six hundred (and) sixty-six and sixty-six hundredths. The negative whole number powers of ten are always read in terms of plural-marked units.

D. Pimm (✉)
Simon Fraser University, Burnaby, BC, Canada
e-mail: djp3@sfu.ca

Is 'six six six' even a spoken number or simply a time-ordered string of digits being listed in turn (one no different from a reading into English of '6, 6, 6' rather than '666'), one that ignores the positional structure? In French, certain numbers (such as phone numbers) whose cardinal value is seldom of interest are frequently read (and written) as sequences of two-digit numbers: 02 65 47 23 46. I will come back to this later too when querying whether the number-word system of any language reflects place value (or better put perhaps, in relation to speech, 'temporal value'). My broader point is that there are significant differences between speaking and writing in relation to numbers, most particularly when it comes to engaging with the written symbolism of mathematics (not least of number), differences that are forgotten at our peril.

As numerical anthropologist Stephen Chrisomalis claims, 'The linkages between number words, computational technologies, and number symbols are complex, and understanding the functions each serves (and does not serve) will help illustrate the range of variability among the cognitive and social systems underlying all mathematics' (Chrisomalis 2009, p. 496).

What Is Written and What Is Said

I start by echoing the claim from early on in Chap. that 'whole number arithmetic is not culture-free, but deeply rooted in local languages and cultures with the inher-ent difficulty of transposition and culture perspective' (this volume, Sect.). As mathematician René Thom once observed:

> when learning to speak, a baby babbles in all the phonemes of all the languages of all the world, but after listening to its mother's replies soon learns to babble in only the phonemes of its mother's language. (cited in Ziman 1978, p. 18)

Also, from Chap. opening page, the expression 'cultures of speaking' brought to mind the fact that there are 'cultures of writing' too (e.g. the order of writing of the two numerals within a single fraction – see Bartolini Bussi et al. 2014) and that these two may not perfectly align within a single 'culture' (see later for a further example involving grouping of digits within a large number in relation to how they are read). And these both influence and are influenced by the physical actions and gestures implicated in counting and computation (a fact worthy of the historical and geographical term 'cultures of gesture', such as varied forms of finger counting and finger calculation[3] – for many examples and a classification scheme, see Bender and Beller 2012). It is important to remember that, in many times and places, these two mathematical actions (counting and computation) were barely connected at all –

[3]One instance is recorded in the writings of the Northumbrian monk Bede (674–723 CE): for example, in *De temporum ratione*. O'Daly (2014) writes: 'The hand, the most portable device of all, was a powerful tool for symbolic representation, calculation, and mental processing in the Middle Ages, and indicates the presence of a comprehensive, but elusive, gestural vocabulary, the full meaning of which we can only guess'.

e.g. the combined but unrelated use of Roman numerals to hold numbers and counting boards with which to calculate (see Tahta 1991; Chrisomalis 2010).

In response to the piece about fraction writing order by Bartolini Bussi et al. (2014), I wrote:

> With fractions written by hand, the composite symbol is produced in a given, temporal order. How might that gestural order relate either to what is said or to how what is said is conventionally written? In English, the first word spoken in time is the numerator: is this so for any language? When a fraction word is written down in English, left to right, the numerator is again the first word to be written. (Ditto the question about other languages.) But when the composite symbol for the fraction is produced, there are variations possible, as their terrific vignettes from China and Burma attest. But both examples point to the arbitrary nature of manual symbol formation (in Hewitt's 1999 use of that word) and to the fact that, once made, the symbol retains (almost) no trace of its making [not least its order]. (Pimm 2014, p. 15)

The many cultures of number are fascinating and intricate, and the particularities of language vis-à-vis time and place, in interaction with computational technological devices (which have existed for at least 5000 years), offer a most worthwhile focus for profound attention. In relation to very recent work concerned with what might be termed 'tangible technological gestures' (see Sinclair and de Freitas 2014, not least in regard to Jackiw and Sinclair 2014), languages themselves at times encode forms of gesturing that have their own transparencies and opacities, their own generalities and idiosyncrasies, all of which form part of the complex symbolic world into which all children are born.

For Wittgenstein, language is, initially but fundamentally, reactive, the word not being the origin:

> The origin and the primitive form of the language game is a reaction; only from this can more complicated forms develop. Language – I want to say – is a refinement, 'im Anfang war die Tat' [in the beginning was the deed]. (Wittgenstein 1937/1976, p. 420)[4]

In relation to the deed of counting, the specific pedagogic language of computational practice (e.g. the English arithmetic metaphor in addition of 'borrowing' and 'paying back') brings with it the possibility that it was at one point literal. One potential example taken from Chap. relates to the suggested link between the medieval Latin expression *reservare in manibus* ('to keep in the hands') and the more contemporary French term à *retenir* ('to keep in mind'). It crossed my mind that the former, in relation to abaci and counting boards, might literally refer to what the hands had to do. Elsewhere, Wittgenstein also commented, 'Remember the impression one gets from good architecture, that it expresses a thought. It makes one want to respond with a gesture' (Wittgenstein 1932, p. 22e). This observation reminded me of the Egyptian hieroglyph for million (𓁨), plausibly a human whole-body gesture at the large size of the number.

Language is not separable from culture nor from gesture (especially not in the context of counting). Gestures perhaps have evolved over a longer period and

4 For much more on this, see Zwicky (1992).

perhaps have left their trace on the language.[5] There is also the possibility of temporal slippage of one system in relation to a development within the other, not dissimilar to those identified by Lakoff and Núñez (2000) with regard to the calculus, whereby the (static) talk is mid-nineteenth century, while associated (dynamic) gestures are more seventeenth century in nature (more fitting to the notion and language of a moving variable, a language that is returning with dynamic geometry environments).

In particular, Raphael Núñez examined the co-production of gestures and speech of Guershon Harel proving a result from real analysis. Núñez observes:

> The study of gesture production and its temporal dynamics is particularly interesting because it reveals aspects of thinking and meaning that are effortless, extremely fast, and lying beyond conscious awareness (therefore not available for introspection). (2009, p. 319)

But it is also true that the gestures are co-produced when counting (and that in certain circumstances *constitute* counting), phenomena that are equally worthy of study as their higher mathematical counterparts. Nevertheless, the focus of this chapter as well as its predecessor is on number language and not number gestures, even though I do not wish to dismiss the latter as epiphenomenal, the way labelling them 'paralinguistic' does.

On Place Value

With regard to place value, one of Chapter's central themes, I have three main observations to make.

First, I would like to consider whether the phenomenon of place value exists solely in relation to written numerals (i.e. written marks, nowadays usually, but not always, employing what are termed Hindu-Arabic numerals[6]) and not in respect of written words or characters from a natural language and whether it also could describe aspects of spoken number words in a natural language as well (or even gestural language – query: *what is the structure of number signs in British, American or Chinese sign language*?). This question reflects my increasing uncertainty as to what place value actually is, as well as echoing Tahta's (1991) informed assertion that place value is appreciably overemphasised in Western mathematics teaching –

---

[5] Numeration systems are among the most linguistically stable systems that exist: the pronunciation split within proto-Indo-European languages into classes labelled *centum* and *satem* (two different words for 'hundred', in Latin and Avestan, respectively) is a surface reflection of this. For an intriguing account of zero, see Rotman (1987).

[6] Chrisomalis (2009) helpfully observes, "I use the term 'western' to refer to the signs 0123456789 instead of 'Arabic' and 'Hindu-Arabic', not to deny that this innovation was borrowed from a Hindu antecedent through an Arabic intermediary, but to avoid confusion with the distinct Indian and Arabic numerical notations used widely to this day. Rendering these latter notations 'invisible' through nomenclature is counterproductive and potentially ethnocentric." (p. 496). In response to this, hereafter I use single quotation marks around 'Hindu-Arabic' in this chapter.

and the discussion of Chinese numeration in Chap. relates to this, when the authors claim, 'The transparency of Chinese [number] names is likely to foster students' understanding for place value' (this volume, Sect.). Though if, as I argue below, place value is simply a convention, then there is a strong question as to whether it is something that is amenable to being 'understood', rather than simply complied with (see Hewitt 1999).

My questioning arose from reading Chap.. The authors claim that traces (which they nicely term 'memories') exist within many spoken numeration systems within natural languages. But these are, at best, ordinal traces, with regard to how number names are said in a conventional order (in English, in decreasing powers of ten, although exceptions like four-and-twenty still exist; in German, the decades are systematically said *after* the units, e.g. 54 is *vier-und-fünfzig*, 'four-and-fifty'). This raised two sub-questions: *does it make sense even to ask whether written (or spoken) natural language numeration systems are or are not place value* and, in regard to written numeration systems that were not place value (e.g. the ancient Egyptian one), *what were their spoken counting systems like*?

In regard to the first question, my (admittedly strong, potentially over-strong) conjecture is no *spoken* language-based numeration system **is** place value (not even Asian ones, which would be the most likely contenders). This is because the structure of how number words are formed ensures that their decimal value is encoded as part of the string, thus changing either the written order (of language-specific symbols on the page) or (temporally) the spoken order in which the various parts of the numeral are said aloud does not alter the combined total. It may go against convention (as 'four-and-twenty' does), but it does not produce a different number. (Of course it is true that simply interchanging the 'six' and the 'seven' in 'sixty-seven' and 'seventy-six' changes the value but that ignores the fact that 'six' is part of 'sixty'.) So possibly place value is solely a phenomenon of written, non-language-based numeration systems, and whichever natural language is used cannot help with this.

My second place value observation, which relates to the first, has particular force because of the particularities and peculiarities of manipulatives such as Dienes blocks (also known as multibase arithmetic blocks – see this volume, Section), which are regularly promulgated as a means to assist with acquiring the concept of place value. See Fig.

It is a commonplace pedagogic move in English-language primary schools to use large sheets of paper and columns labelled (from left to right), thousands (or Th), hundreds (or H), tens (or T) and units (or U).[7] The blocks are collected and placed in the respective columns and then 'Hindu-Arabic' digits are used to record the number of them in each column, hugely finessing the fact that it is actually the paper columns and not the blocks themselves that are both 'holding' the places and, consequently, carrying 'place value'.

---

[7]And these too can go on, TThs for tens of thousands, HThs for hundreds of thousands, Ms for millions, etc. Note how both words for TThs and HThs are plural, but in the notation, only the latter is marked symbolically. For more on this.

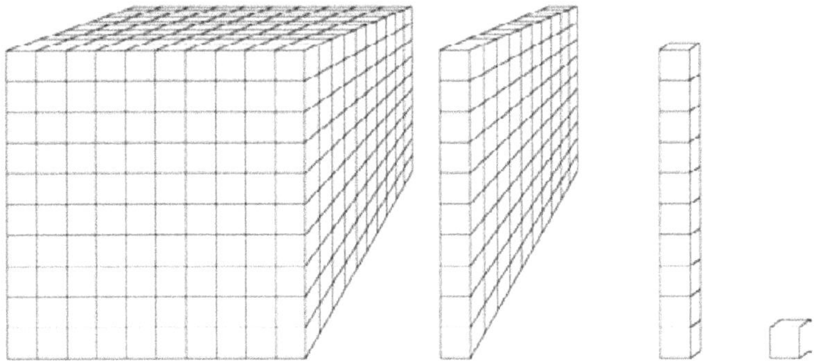

Fig. Dienes blocks (1000, 100, 10 and 1)

However, as with the instance mentioned in Chap. where a 7-year-old writes 10013 instead of 113, an accurate (as opposed to conventionally correct) notating from a corresponding Dienes block paper configuration would be 1H, 1T, 3Us = 100, 10, 3 = 100103. The question of quantity versus place is an intricate and arbitrary one (again in Hewitt's 1999 sense), and there is no *necessary* reason why there cannot be any number of blocks of any size in a single column (something an abacus masks by each spike having a set, uniform height relative to the diameter of the beads). Indeed, a higher (and linear-algebra-influenced) mathematical perspective has any whole number generated by the basis consisting of powers of 10 (and includes decimal fractions, if negative powers of 10 are permitted) with the coefficients 0–9. It is partly for this reason that I mentioned the six, six, six reading of '666' in the opening paragraphs of this commentary (as well as linking to David Fowler's 1987 historical reconstruction, via the arithmetic process of *anthyphairesis*, of a pre-Euclidean functioning definition of ratio).

Before raising further difficulties, there are three more observations I wish to make about Dienes blocks themselves. The first is that they can actually be modified to display any whole number or decimal fraction. To have ten thousand, for instance, one simply needs to stick together ten of the large cube size; for hundred thousand, a square array of a hundred of the large cube size; and for million, a cube array of a thousand the large cube size. For decimal fractions, merely rename one of the larger blocks (e.g. the larger cube or the 'long' or the 'flat' as they are sometime called) as 'one'.[8] Secondly, this visuo-geometric repetition every 10^3 exactly fits the SI (*Système Internationale*) emphasis on grouping whole number digits into triples, as well as reflecting the standard metric naming structure of measures (though if we

[8] As is so often the way of these things, having figured this out for myself, I then came across the paper by Kim and Albert (2014) which purports both to give an account of and to account for the history of base-ten blocks. While disagreeing profoundly with much of their accounting for, they did remind me that in Dienes' (1963) book *An Experimental Study of Mathematics Learning* – a book which I read in 1972 as part of a very early mathematics education course taught by David Tall in the University of Warwick mathematics department – Dienes makes the same observation, on his p. 28.

wanted to use this to refer to a kilo-something, a milli-something or a micro-something, we would need a word for a standard counting unit – other than 'unit'). My third observation is straightforward: there is a 'natural' and directly observable sense of decimal equivalence between each power of ten and the next one.

However, it strikes me that Dienes blocks are at least as good a fit to the base-ten system of Egyptian hieroglyphic numerals (which uses the repetition of vertical lines, hoops, scrolls, lotus flowers, etc. where there are no links whatsoever among the symbols for 1, 10, 100, 1000 and so on, either to record numbers or to calculate with them).[9] (See Fig.) And this numeration system is decidedly *not* a place value one.

In general, there are two alternate principles for generating words or other symbols for numbers: repetition (related to tallying) and cypherisation, namely, the use of distinct and independent symbols for each number. Many older symbolic systems use repetition as the primary principle. For example, below (see Fig.) depicts an example using the ancient Egyptian numeration system: on the left is the conven-tional order and on the right in a scrambled order. (Whole-number adding is totally unproblematic as the symbols themselves are simply combined, and any excess over nine of one power of ten is converted into one of the next power up.)

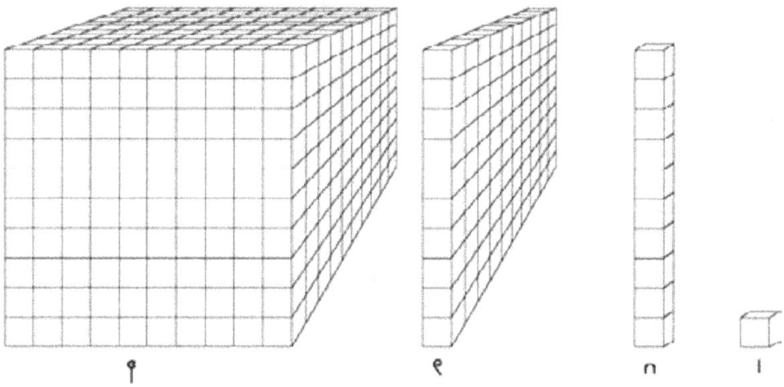

Fig. A depiction of Dienes blocks (with ancient Egyptian hieroglyphs for 1000, 100, 10 and 1 appended)

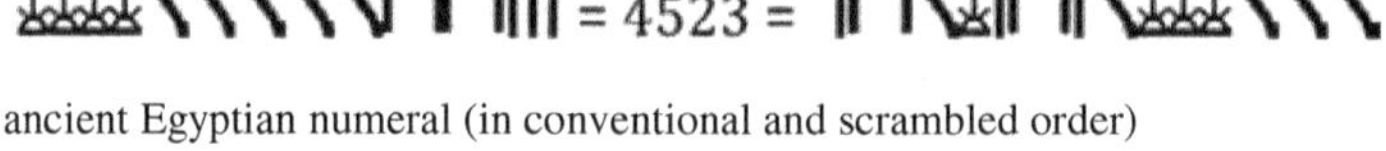

Fig. An ancient Egyptian numeral (in conventional and scrambled order)

[9]This is not always the case historically. With Roman numerals, for instance, they are perfectly competent for recording numbers. However, due to a variety of principles being combined (both additive and subtractive, the use of 5, 50, 500, etc. as an intermediate resting place sometimes called a sub-base, etc.), they are not easily used for calculation (particularly multiplication and division). But the companion use of counting boards was perfectly adequate for that.

'Full' cypherisation means every initial numeral up to one less than the base has a different symbol (think 1, 2, 3, 4, 5, 6, 7, 8, 9). Chinese rod numerals (discussed in Chap.) reflect enormous flexibility with very limited cypherisation (the vertical rod and the horizontal rod), akin to Roman numerals I and V, only without the latter's subtractive principle[10] and much repetition, likely because they became traces of actual piles of rods used on counting boards, where an attribute of the board (lines, positions) took care of the place value. Likewise, ancient Babylonian numeration consists in its entirety of only two distinct stylus edge marks (one the same as the other only rotated through 90°), together with a mixed-base system (ten and sixty), repetition, a form of place value and contextual 'floating point'.

One pragmatic test for any written system (presuming it makes repeated use of the same set of symbols or objects) with regard to its being place value or not is whether one can generically scramble the order of the marks and not affect the numerical value represented: this is true with Dienes blocks and also with ancient Egyptian numerals. In passing, it is also true of early Greek (Ionian) numerals (and are still used today for depicting ordinals), where the numeral for 1 bears no relation to that of 10 or 100, being different letters of the alphabet. The Egyptian numerals (as do the Greek) retain their specific decimal value, even when rearranged – see Pimm (1995) for more on these complexities of symbol/object manipulation. Consequently, Dienes blocks cannot 'contain' place value. So, if they do 'work', how do they 'work'? The 'value' is there, but the 'place' is not.

Caleb Gattegno (e.g. Gattegno 1974) repeatedly proposed systematising language- based counting systems in elementary schools in different European languages, in order to make them easier to learn by being a far closer fit to the standard Western written numeration system. In particular, in English he wanted *ten* to be said as 'one-ty', *eleven* as 'one-ty-one' and *twelve* as 'one-ty-two'[11] and then *twenty* as 'two-ty', *thirty* as 'three-ty' and so on. With the later decades (*sixty*, *seventy*, *eighty*, *ninety*), the changes merge with the actual empirical system. There is a *samizdat*-style community within Anglophone mathematics education (especially within the UK, but in North America and Western Europe too), which works extensively with Gattegno's ideas (and the teaching aid of the Gattegno number chart, among others), in order to support the acquisition of structured fluency in number naming (for one recent instance, see Coles 2014).[12] The Chinese numeration system detailed in Chap. (which presents in several other Asian languages too) has these properties already.

My third place value observation relates to the notion of linguistic/conceptual transparency as employed in Chap. , the generation of number terms for powers of

[10] For example, instead of IIII, they used IV = V – 1; instead of XXXX, they used XL = L – X, but this shortening only occurred in mediaeval times.

[11] The etymology of *eleven* and *twelve* is quite singular, deriving from old Norse words 'einlief' and 'twalief', meaning, respectively, 'one-left' and 'two-left' (presumably after taking away ten, a trace subtractive principle that can also be seen in Roman numeration).

[12] There is also a far greater emphasis in this community on the value of acquiring ordinal elements of counting: see Tahta (1998) or Coles and Sinclair (2017).

ten and how they relate to the standard (SI) means of writing large whole numbers using 'Hindu-Arabic' numerals. In Chinese, 千 (*qiān*[13]) is the character for 'thousand' although none is needed, as it is not in English either, based on the principle that a new power-of-ten name is only needed when the same two terms would otherwise be next to each other. 'Ten tens' gives rise to 'hundred', but ten hundreds ('thousand') should cause no difficulty (and does not in naming centuries, e.g. 'the seventeen hundreds') and need not exist, while 'hundred hundreds' is the next one that should generate a new term. In Chinese, that character is 万 (*wàn*), while English speakers just say 'ten thousand'. (To belabour the point, notice it is not said as 'ten thousand**s**', as conventional pluralisation rules of English would demand – though see the next section on the distinction between mass and count nouns.[14]) It is this same principled issue that causes divergent interpretations of 'billion' ('hundred million' in North America, 'million million' in the UK, at least historically): likewise with 'trillion'. But the generation of these new words for certain powers of ten allows the use of the same number words from one to nine to be combined to name every whole number.

The SI number convention declares:

> The digits of numerical values having more than four digits on either side of the decimal marker are separated into groups of three using a thin, fixed space counting from both the left and right of the decimal marker. Commas are not used to separate digits into groups of three. (http://physics.nist.gov/cuu/Units/cheklist.html)

Thus, for example, 213 154 163 is how this number *should* be written. However, this convention makes a (false, universal) presumption in relation to every natural language on the planet with regard to the structure of number words within each language, because, as I mentioned at the outset, written numerals are not part of any natural language.

So in relation to transparency and Chapter's claimed 'perfect' match of Chinese numeration and 'the mathematician's arithmetic', this is one place where the Chinese language numeration system does not match SI at least, namely, with regard to delimiting (whether by means of commas, full stops or spaces) numerals with more than four digits. For instance, twelve thousand is written in Chinese as 一万两千 (*yīwàn liǎngqiān*; in other words, one wàn two qiān – one ten-thousand and two thousand), which does not match 12 000. The written symbolic form of numbers can thus aid or interfere with generating the corresponding correct, language- specific spoken form.

In an article about this particular mismatch, Arthur Powell opens his account as follows:

> In May and June of 1984, while conducting a series of mathematics teacher education workshops in Beijing, capital of the People's Republic of China, I was introduced to some pedagogical problems in Chinese numeration. They involve the teaching and learning of

[13] Throughout this chapter, and indeed this volume, Hanyu Pinyin (a standardised Romanisation system for Mandarin Chinese) is used to represent Chinese characters.

[14] The Greek word for ten thousand – *myriad* – is used in English as a 'round' number word for a very large number (for much more on the linguistics of round numbers, see Channell 1994).

> how to speak numerals with fluency in Chinese, using Hindu-Arabic written numerals. A salient feature of these problems manifests itself when Chinese students attempt to read numerals longer than four digits. For example, even graduates of senior middle schools find it necessary to read 6,721,394 by first pointing at and naming from right to left the place value of each digit before knowing how to read the "6" in the millionth place and the rest of the numeral. (Powell 1986, p. 20)

Powell's proposals with regard to a way of ameliorating this difficulty in this article relate to a suggestion generated by Gattegno's ideas of the power of pedagogic modification of certain elements of number-naming systems in order to emphasise structure:

> [this proposed alternative approach] allows learners to become aware of the regularity of Chinese numeration. It also helps learners to develop strategies for by-passing reading difficulties caused by the employment of a convention of delimiting digits which is contradictory to the linguistic structure of Chinese. (p. 20)

So, by putting a space or comma after every four digits (rather than three) and reading the delimiter as *wàn*, correctly spoken Mandarin Chinese numeration follows, rather than it having to be memorised.[15] But this does raise the question of where, in regard to mathematics, does a specific language 'stop'.

Finally, I was led to wonder, *if place value can be so transparent in some systems, whether it becomes hard to think about change of base.* But I cannot go into this here. The next core element of this commentary relates to the complex issue of numerical units.

Count Nouns and Mass Nouns: The Question of Units

The motivation for exploring the issues of this section arose in part from the interesting and important discussion of number classifiers in Chinese provided in Chap. , but also from my simple curiosity wondering why in lots of settings English number words have features of nouns that reflect both singular and plural forms: for instance, in the everyday expression 'hundreds and hundreds' compared with 'two hundred (and) fifty-three'. Or, the spoken number following 'ninety-nine' is 'one hundred' or 'a hundred', yet the number following one hundred ninety-nine is 'two hundred'). Why is it that 'hundred', when used as a power-of-ten unit, is singular (e.g. two hundred and forty-two, rather than two-hundred**s** and forty-two)? Why do number words put pressure on the straightforward singular/plural distinction in English? How does this play out in the units for the countable noun and are numbers themselves such units? As Wittgenstein observed, 'Grammar tells what kind of

[15] To illustrate how devastating this can be, I noticed today that my ticket for the Vancouver SkyTrain has a twenty-digit identification number, presented in groups of four: 0001 1570 5839 8568 8326. Had I to read this aloud as a single number (were I concerned, say, about its cardinality), rather than as simply a string of digits with a pause for each space, I would be rendered speechless.

object anything is' (Wittgenstein 1953, p. 116); hence, this uncertainty (present also in singular or plural verbs) potentially reflects an ontological instability at the heart of (English-language) number.

The earlier discussion of Dienes blocks and the paper tabular presentation also had column headings that were called 'thousands', 'hundreds', 'tens' and 'units'. Yet once numerals were used to replace the (multiple) blocks, the differentially marked plural forms vanished in both the corresponding spoken and written English. (With regard to fractions words in English, not least the question of whether 'three-fifths' is a singular or plural noun and how it differs syntactically from 'three fifths', see the next section.)

In order to pursue some of the challenges that were mentioned in the previous section, I wish to examine certain morphosyntactic aspects of number words in English. One broad distinction in English grammar (which has commented upon in the literature since at least the early 1900s[16]) is the distinction between count nouns and mass nouns (the latter is sometimes referred to as 'non-count' nouns, though the two categories are not the same – see Laycock 2010), albeit one currently eroding (as is the case with round numbers) in interesting ways.

Edward Wisniewski begins his chapter on the potential cognitive basis for such a distinction as follows:

> English and other languages make a grammatical distinction between count nouns and mass nouns. For example, "dog" is primarily used as a count noun, and "mud" is primarily used as a mass noun. Count nouns but not mass nouns can be pluralized and preceded by numerals (as in "three dogs" but not "three muds"). Count nouns but not mass nouns can appear with the indefinite determiner "a" (as in "A dog ate the chicken" but not "A mud covered the chicken"). On the other hand, mass nouns can appear with indefinite quantifiers, such as "much" or "little" (as in "much mud" but not "much dog"), whereas count nouns can appear with indefinite quantifiers such as "many" and "few" (as in "many dogs" but not "many muds").[17] (2010, p. 166)

There are many things to be said about this distinction. One key observation concerns the potential that any English noun, in certain circumstances, can come to take on both count and mass aspects, rather than, as suggested above, that these are two disjoint noun categories. This is also marked by the failing distinction between *fewer* (count) or *less* (mass), likewise *many* and *much*. In a mathematics context, this flexibility can be seen in the nineteenth century with the terms 'algebra' or 'geometry', where mathematical developments (non-Euclidean geometry, Boolean algebra) subsequently enabled 'an algebra' or 'two geometries' to be spoken of (and related to highly significant shifts in the perception of the underlying mathematics being referred to). In the late twentieth century, 'technology' has morphed to allow

---

[16] This distinction is certainly not limited to English but is far from universally employed across languages. In particular, the various forms of Chinese do not distinguish these categories, instead using classifiers as described in Chap.

[17] There is a footnote in Wisniewski's chapter at this precise point, which begins, 'Some languages such as classifier languages (e.g. Japanese) do not make a distinction between count and mass nouns. Nevertheless, they do have mechanisms for indicating that an entity is or is not individuated'. As explored at length in Chap., Chinese is also a classifier language.

'a technology' or 'digital technologies'. All instances of conventionally mass nouns admit count possibilities and characteristics, and vice versa.[18]

A second observation has to do with the way that mass nouns are quantified: traditionally, this was by using various instances of the 'a unit of' construction (e.g. 'a slice of' or 'a loaf of' bread, 'a grain of' rice), where the unit could always be quantified, i.e. was itself a count noun. (Though, contrariwise, seeing four as 'a quartet of'[19] permits all count nouns to be seen as mass nouns – albeit in the plural form – where the number words themselves can be quantified: two quartets of, three septets of, and so on.[20])

However, the most educationally significant thing by far that this distinction relates to in regard to this chapter and the previous one is the extent to which number words themselves (in English or other languages), when functioning as nouns (as they do in arithmetic), do so as mass or count nouns. This is centrally related to the passing comments I have repeatedly made so far about whether it should be 'hundred' or 'hundreds'.

In the specific context of this chapter, however, my interest lies with English number words themselves in their nominal form, one, two, three, etc., and the corresponding ordinals, first, second, third (which may or may not function as nouns), and the somewhat bewildering connection in some languages between ordinals and fraction terms (see Pimm and Sinclair 2015, as well as the next section). With all of these sets of number words, the question is: mass or count?

Consider the English word count sequence 'one, two, three, etc.'. One of the ambiguities in English in respect of multiplication has to do with whether it 'should be' four twos *is* or four twos *are* eight.[21] Notice the distinct pluralisation of 'two' marks it as a count noun, as does the 'count' word 'four', as does the verb agreement of 'are' with the pluralised noun form 'twos'. The presence of a count noun permits the question 'How many?' to be asked in relation to it (for much more on this, see Sinclair and Pimm 2015a). Yet in the count sequence one, two, three, etc., the number words act more like mass nouns. And, as always, *what cognitive shifts or chasms underlie such linguistic uncertainties*?

Look at the ordinal terms: first, second, third, fourth, etc. While it is possible to imagine scenes where a count noun perspective is possible (e.g. in an athletic meeting, asking a runner: how many firsts, and how many seconds?, meaning first places and second places), these act more like mass than count nouns. But notice what happens when we shift to the related fraction forms: again, we get two sevenths and

---

[18] As a potential example of nouns going the other way, consider 'I returned to the car and there was bird all over the windshield', though this also could be seen as an anti-synecdoche, using the whole for the part.

[19] With regard to Ancient Greece, David Fowler refers to *arithmoi* as cardinals, but helpfully observes, 'a much more faithful impression of the very concrete sense of the Greek *arithmoi* is given by the sequence: duet, trio, quartet, quintet, …' (Fowler 1987, p. 14).

[20] Cf T. S. Eliot's *Four Quartets* as well as the earlier mention of Fowler and *arithmoi*, specifically in the previous footnote.

[21] Though here the question, in relation to the calculator and what is said when pressing the '=' button, is perhaps whether the verb should not be 'is (are)' but rather 'make(s)' or 'give(s)'.

three tenths, which are plural and count nouns (and would encourage this syntax for statements like 'two sevenths *are* bigger than three tenths'). But then the unifying hyphen may show up (two-sevenths, three-tenths), with the effect of singularising these composite forms.[22]

Once again, this section simply contains some brief comments and observations about number language in certain contexts. In the next section, I move towards aspects of a range of number word systems.

Cardinal, Ordinal and Fractional: Three Interlocking Linguistic Subsystems

Whole numbers are not the only game in town. Languages also have systematic ways of naming (of summoning, of calling into being) ordinals and fractions (decimals or otherwise) as well. Most of what follows specifically concerns the English language, though a more diverse discussion (concerning some twenty different languages) can be found in Pimm and Sinclair (2015), which explored variations among these three sets of number words across a variety of languages and language groups. The motivation to do so arose from the two papers in *For the Learning of Mathematics* (Bartolini Bussi et al. 2014; Pimm 2014), most specifically in relation to close links (in some languages near identical) between how ordinal words and fraction words are formed (and why this might be).

In an attempt to summarise some of what was found, here are four diagrams that reflect different relationships among cardinal (C), ordinal (O) and fraction (F) words within specific languages from my dataset. The arrows indicate 'adding' a suffix to the previous sets of words to form the new set. Figure (a) captures, e.g. Norwegian, while (b) exemplifies one common relationship (e.g. German): (c) is the 'degener-ate' case of (b) that fits some Western European languages (e.g. English, French, Italian and Spanish), while (d) reflects Hungarian.

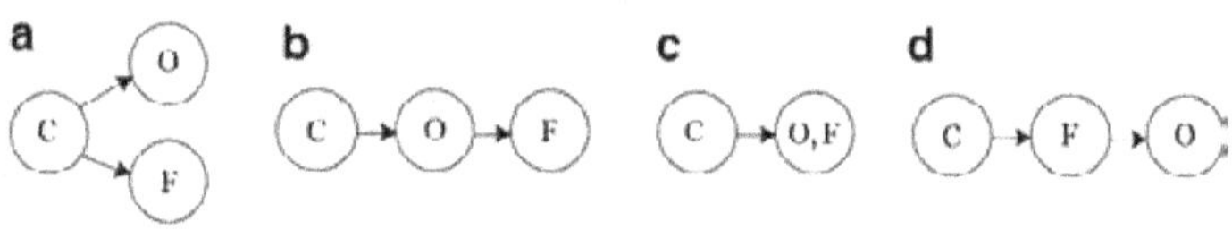

Fig. (a–d) Various relationships among sets of number words within a single natural language

---

[22] Pedagogically, Hewitt (2001, pp. 47–8) explores fraction operations based primarily on linguistic parallels between invented non-number noun names (that he gives compound number name patterns to, such as 'flinkerty-floo' or 'zipperly-bond') and number nouns, asking not only 'how many twenty-fourths are there in one?' and 'how many four-hundred-and-twentieths are there in nine?' but also 'how many flinkerty-flooths are there in one?' and 'how many flinkerty-flooths are there in zipperly-bond?'. Note his use of 'How many?' questions and plural number nouns and plural verbs throughout.

As I mentioned at the outset, one question I was led to examine was how these naming systems relate one to another within a given language, as well as how they relate to gestures on the one hand and to trans-linguistic written numerals on the other. Ordinality primarily concerns the sequential aspect of whole numbers, as opposed to their quantitative (cardinal) one. And there is a key and fundamental question about which came first, cardinal or ordinal. (For much more on this, see Seidenberg 1962 and Sinclair and Pimm 2015a.) But spoken ordinal terms carry a significant difference from cardinal terms in that the core issue becomes which one comes before or is said before or after another, rather than which one is bigger or smaller. Thus, ordinality is strongly related to temporality rather than magnitude.

Here are two minute observations. First, there is a commonly employed, hybrid written notation that seems both to pull 'Hindu-Arabic' numerals into a specific language and to privilege the cardinal over the ordinal: 1st, 2nd, 3rd, 4th, 5th, etc. (even though in French it is 1er, 2ième, 3ième, 4ième, 5ième, etc.). The second one is specific to English and relates to the supposed cardinal counting decade words: both 'thirty' and 'fifty' show explicit ordinal over cardinal traces – 'thir-ty' as the third '-ty', 'fif-ty' as the fifth '-ty' – a visible (and audible) trace, not least because of the distinction between the English words 'three' and 'third' and 'five' and 'fifth' (from the two closely related English language systems of cardinal and ordinal words), whereas neither 'four' and 'fourth' exactly 'fit' 'forty'.

There is an ordinal regularity in English after five, both of forming ordinals 'from' cardinals and the presumed economy (and greater ease of pronunciation) of potentially dropping the '-th' suffix from a possible, historical *sixthty*, *seventhty*, *eighthty* and *ninethty*. But in regard to my discussion in Pimm (2014) of '*the* fifth part' (in regard to the singularity of unit fractions in Ancient Egyptian arithmetic), there is some appreciable specificity implied, in that 'the sixthty' (seen as 'the sixth '-ty'') would have to be unique and 'a sixthty' or 'two sixthtys' would not be feasible.

A Few Concluding Remarks

The main focus of this commentary piece has been to draw attention to certain features of number language, both language-specific ones (mostly in regard to English) and across certain classes of language (in terms of the presence or absence of certain distinctions, such as mass/count or classifiers) that may have some pertinence or significance in learning to number and to count. But underneath it has been an attempt to keep an ear and an eye out for 'traces' ('memories' in the terminology of Chap.) of what has passed before or en route (both within individuals and within cultures) to our present-day set of practices and forms with regard to number.

In particular, in my attempt to localise place value away from natural language and primarily into written symbolic notation systems (though it is important not to forget physical manifestations of the same, such as with *khipu* – see Chrisomalis

2009), I have endeavoured to make distinctions among the interlinked systems of language, notation and the world. In regard to mathematics education, far more generally, the potential overvaluing of cardinal number as the pedagogically presumed dominant form with regard to arithmetic and mathematics has some serious consequences, as has the consequent downplaying of ordinality and its significant role in learning how to count (see Tahta 1991, 1998; Sinclair and Pimm 2015a, b; Coles 2017).

In Fig., there is my first attempt at trying to depict this (even though I already can see problems, oversimplifications, omissions and errors). It draws on the distinction between metaphoric and metonymic relations, as outlined in Tahta (1991, 1998), which he links with the abacist and the algorist, respectively: the use of physical objects (which become metaphors for number) versus the 'manipulation' of numerals.

Tahta writes:

> Metaphor and metonymy are not necessarily distinct polarities, but more like aspects that can be stressed or ignored as desired. One of our problems in teaching arithmetic is the move from the stress on metaphor to the stress on metonymy. We offer children counters and rods and so on, in order to mimic processes which we eventually want them to transfer to written or spoken numerals. (1998, p. 6)

As an individual becomes more and more numerically fluent, the separation between number words and numerals becomes less and less: but this does not mean that those distinctions and separations cease to leave their traces.

One final observation: Chrisomalis's (2010) fascinating book on the history of numerical notation is over five hundred pages long. The world and its (linguistic) history in regard to whole number is a very complex and sophisticated mix. But also an engaging and, at times, fascinating one.

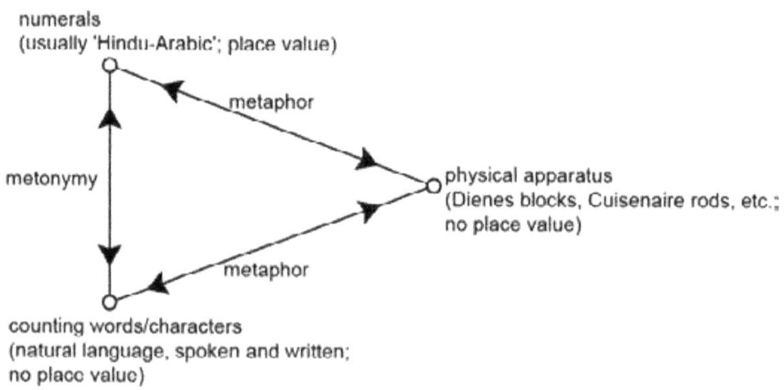

Fig. Metaphor and metonymy in relation to the interlinked systems of natural language, notation and the world

References

Bartolini Bussi, M., Baccaglini-Frank, A., & Ramploud, A. (2014). Intercultural dialogue and the geography and history of thought. *For the Learning of Mathematics, 34*(1), 31–33.

Bender, A., & Beller, S. (2012). Nature and culture of finger counting: Diversity and representational effects of an embodied cognitive tool. *Cognition, 124*, 156–182.

Channell, J. (1994). *Vague language*. Oxford: Oxford University Press.

Chrisomalis, S. (2009). The cognitive and cultural foundations of numbers. In E. Robson & J. Stedall (Eds.), *The Oxford handbook of the history of mathematics* (pp. 495–517). Oxford: Oxford University Press.

Chrisomalis, S. (2010). *Numerical notation: A comparative history*. Cambridge: Cambridge University Press.

Coles, A. (2014). Transitional devices. *For the Learning of Mathematics, 34*(2), 24–30.

Coles, A. (2017). A relational view of mathematical concepts. In E. de Freitas, N. Sinclair, & A. Coles (Eds.), *What is a mathematical concept?* (pp. 205–222). Cambridge: Cambridge University Press.

Coles, A., & Sinclair, N. (2017). Re-thinking place value: From metaphor to metonym. *For the Learning of Mathematics, 37*(1), 3–8.

Dienes, Z. (1963). *An experimental study of mathematics learning*. London: Hutchinson.

Fowler, D. (1987). *The mathematics of Plato's academy: A new reconstruction*. Oxford: Oxford University Press.

Gattegno, C. (1974). *The common sense of teaching mathematics*. New York: Educational Solutions.

Hewitt, D. (1999). Arbitrary and necessary part 1: A way of viewing the mathematics curriculum. *For the Learning of Mathematics, 19*(3), 2–9.

Hewitt, D. (2001). Arbitrary and necessary: Part 3 educating awareness. *For the Learning of Mathematics, 21*(2), 37–49.

Jackiw, N. & Sinclair, N. (2014). *Touch counts* [software application for the iPad]. https://itunes.apple.com/ca/app/touchcounts/id897302197?mt=8

Kim, R., & Albert, L. (2014). The history of base-ten-blocks: Why and who made base-ten-blocks? *Mediterranean Journal of Social Sciences, 5*(9), 356–365.

Lakoff, G., & Núñez, R. (2000). *Where mathematics comes from: How the embodied mind brings mathematics into being*. New York: Basic Books.

Laycock, H. (2010). Mass nouns, count nouns, and non-count nouns: Philosophical aspects. In A. Barber & R. Stainton (Eds.), *Concise encyclopedia of philosophy of language and linguistics* (pp. 417–421). Amsterdam: Elsevier.

Núñez, R. (2009). Gesture, inscriptions, and abstraction: The embodied nature of mathematics or why mathematics education shouldn't leave the mathematics untouched. In W.-M. Roth (Ed.), *Mathematical representation and the interface of body and culture* (pp. 309–328). Charlotte, NC: Information Age Publishing.

O'Daly, I. (2014). *Talk to the hand: Finger counting and hand diagrams in the Middle Ages*. https://medievalfragments.wordpress.com/2014/03/14/talk-to-the-hand-finger-counting-and-hand-diagrams-in-the-middle-ages/

Pimm, D. (1987). *Speaking mathematically: Communication in mathematics classrooms*. London: Routledge and Kegan Paul. (To be reissued in 2017.)

Pimm, D. (1995). *Symbols and meanings in school mathematics*. London: Routledge.

Pimm, D. (2014). Unthought knowns. *For the Learning of Mathematics, 34*(3), 15–16.

Powell, A. (1986). Economizing learning: The teaching of numeration in Chinese. *For the Learning of Mathematics, 6*(3), 20–23.

Rotman, B. (1987). *Signifying nothing: The semiotics of zero*. London: Macmillan.

Seidenberg, A. (1962). The ritual origin of counting. *Archive for History of Exact Sciences, 2*(1), 1–40.

Setati, M. (1998). Code-switching in a senior primary class of second-language mathematics learners. *For the Learning of Mathematics, 18*(1), 34–40.

Sinclair, N., & de Freitas, E. (2014). The haptic nature of gesture: Rethinking gesture with new multitouch digital technologies. *Gesture, 14*(3), 351–374.

Sinclair, N., & Pimm, D. (2015a). Whatever be their number: Counting on the visible, the audible, and the tangible. In M. Meletiou-Mavrotheris, K. Mavrou, & E. Paparistodemou (Eds.), *Integrating touch-enabled and mobile devices into contemporary mathematics education* (pp. 50–80). Hershey, PA: IGI Global.

Sinclair, N., & Pimm, D. (2015b). Mathematics using multiple senses: Developing finger gnosis with three- and four-year-olds in an era of multi-touch technologies. *Asia-Pacific Journal of Research in Early Childhood Education, 9*(3), 99–110.

Tahta, D. (1991). Understanding and desire. In D. Pimm & E. Love (Eds.), *Teaching and learning school mathematics* (pp. 221–246). London: Hodder and Stoughton.

Tahta, D. (1998). Counting counts. *Mathematics Teaching, 163*, 4–11.

Wisniewski, E. (2010). On using count nouns, mass nouns, and *pluralia tantum*: What counts? In F. Pelletier (Ed.), *Kinds, things, and stuff: Mass terms and generics* (pp. 166–190). Oxford: Oxford University Press.

Wittgenstein, L. (1932). *Culture and value* (P. Winch, Trans.). Chicago: University of Chicago Press.

Wittgenstein, L. (1937/1976). Cause and effect: Intuitive awareness. *Philosophia, 6*(3–4), 409–425.

Wittgenstein, L. (1953). *Philosophical investigations* (G. Anscombe, Trans.), Oxford: Blackwell.

Ziman, J. (1978). *Reliable knowledge: An exploration of the grounds for belief in science.* Cambridge: Cambridge University Press.

Zwicky, J. (1992). *Lyric philosophy*. Toronto: University of Toronto Press.

Cited papers from Sun, X., Kaur, B., & Novotna, J. (Eds.). (2015). Conference proceedings of the ICMI study 23: Primary mathematics study on whole numbers. Retrieved February 10, 2016, from www.umac.mo/fed/ICMI23/doc/Proceedings_ICMI_STUDY_23_final.pdf

Pimm, D., & Sinclair, N. (2015). The ordinal and the fractional: Some remarks on a trans-linguistic study (pp. 354–361).

5

The What and Why of Whole Number Arithmetic: Foundational Ideas from History, Language and Societal Changes

Xu Hua Sun, Christine Chambris, Judy Sayers, Man Keung Siu, Jason Cooper, Jean-Luc Dorier, Sarah Inés González de Lora Sued, Eva Thanheiser, Nadia Azrou, Lynn McGarvey, Catherine Houdement, and Lisser Rye Ejersbo

Introduction

Mathematics learning and teaching are deeply embedded in history, language and culture (e.g. Barton 2008). Yet what historical, linguistic and cultural foundations are necessary for the early years of school to adequately prepare children for mathematics learning? To address this question, we summarise work on these three aspects of WNA to frame the entire volume and identify the historical, linguistic and cultural bases on which other aspects of learning, teaching and assessment are based. The chapter provides a meta-level analysis and synthesis of what is known about WNA's foundations of history, language and societal changes, which serves as a useful base from which to gauge any gaps and omissions. This foundation also

The first author was supported by the Research Committee, University of Macau, Macao, China (MYRG2015-00203-FED). The opinions expressed herein are those of the authors.

The first two authors wrote the chapter with the collaboration of other working group members.

X.H. Sun (✉)
Faculty of Education, University of Macau, Macao, China
e-mail: xhsun@umac.mo

C. Chambris (✉)
Laboratoire de didactique André Revuz, Université de Cergy-Pontoise, Cergy, France
e-mail: christine.chambris@u-cergy.fr

J. Sayers
Stockholm University, Stockholm, Sweden

M.K. Siu
University of Hong Kong, Pokfulam, Hong Kong, China

provides an opportunity to learn from the practices of different times and languages and from societal changes.

Conference Presentations: Overview

Thirteen papers written by authors from 11 countries were presented for Theme 1. For presentation and discussion, these papers were divided into four subgroups exploring several overlapping aspects of the why and what of WNA: the historic background of WNA, the language foundations of WNA, the foundational ideas that underlie WNA and the support for societal changes to the teaching and learning of WNA.

Historical Background

Zou (2015) summarised findings from historical investigations of arithmetic in ancient China, including how number units were derived and named and how numbers were represented with rod or bead calculation tools and with symbols. Siu (2015) studied the book of *Tongwen Suanzhi* (同文算指) (Rules of Arithmetic Common to Cultures, 1614) and reviewed how counting rods and the abacus were gradually replaced with written calculations in China. Sun (2015), also discussing early Chinese development, presented the use of advanced number names and

J. Cooper
Weizmann Institute of Science, Rehovot, Israel

J.-L. Dorier
University of Geneva, Geneva, Switzerland

S.I.G. de Lora Sued
Pontificia Universidad Católica Madre, Santiago de los Caballeros, República Dominicana

E. Thanheiser
Portland State University, Portland, OR, USA

N. Azrou
University Yahia Fares Medea, Medea, Algeria

L. McGarvey
University of Alberta, Edmonton, AB, Canada

C. Houdement
Laboratoire de didactique André Revuz, Université de Rouen-Normandie, Rouen, France

L.R. Ejersbo
Aarhus University, Aarhus, Denmark

calculation tools (counting rods and the *suàn pán* or Chinese abacus) and emphasised how place value is still the most overarching principle of WNA based on Chinese linguistic habit. Traces of this influence can still be found in contemporary core curriculum practices in many countries today.

Language Foundation of WNA: Regularity, Grammar and Cultural Identity

Azrou (2015) reported how the historical and linguistic colonisation of Algeria affected the learning of WNA and presented the first step of an intervention for teacher education, which can also promote students' cultural identities. Chambris (2015) showed how changes related to place value that were introduced by the New Math in France (1955–1975) continue to be influential today.

Houdement and Tempier (2015) reported on two experiments for strengthening the decimal (base ten) principle of numeration, assigning a key role to the use of numeration units in France. Changsri (2015) explored first grade students' ideas of addition in two Thai schools in the context of lesson study and an open approach and found that the students used a variety of representations to express addition ideas.

Foundational Ideas Underlying WNA

Dorier (2015) gave an overview of the development of numbers, showing how Brousseau's theory can be used in accordance with this historical context to develop the key stages of a teaching sequence using the concept of numbers. Thanheiser (2015), also studying teacher education, adopted the perspective of variation theory and used historical number systems as a tool, finding that prospective teachers developed a more sophisticated concept of the base-ten place value system by examining, comparing and contrasting different aspects of historical systems. Ejersbo and Misfeldt (2015) described research introducing a regular set of number names in primary schools in Denmark. Sayers and Andrews (2015) summarised an eight-dimensional framework called foundational number sense (FoNS) that characterises necessary learning experiences for young children. They demonstrated how to use the framework by analysing learning opportunities in first grade in five European contexts.

Different Expected Learning and Teaching Goals for WNA

Cooper (2015) discussed how a university mathematician and a group of elementary school teachers, working together in a professional development course, revealed new insights into division with remainders. McGarvey and McFeetors (2015) identified the Canadian public's concerns about the goals of WNA and the support required for students to reach them.

Working Groups' Discussions

The eight one-hour sessions were organised in different ways. Examining variation in WNA across history and language and across different communities, working group 1 discussed the implications of different views on the why and what of WNA for instruction and teacher education. Place value in the so-called Hindu-Arabic system was discussed extensively in the working group sessions. In addition to the background discussion and questions posed in the Discussion Document (this volume, Appendix 2), the papers for Theme 1 can facilitate discussions of the following questions:

1. How has the place value concept developed across numeral systems?
2. What are the issues of language and culture in WNA?
3. How did/do different communities change past/current teaching of WNA?

The Structure of This Chapter

This chapter describes number representations and their foundational ideas beyond the variations in WNA across history, language and culture. As the world becomes more unified and previously separate cultures interact and begin to merge, incompatibilities become visible and separations arise between different traditions and practices. Many of these incompatibilities can be seen in WNA. Particularly notable is the accommodation needed when traditional language is adapted to deal with the nearly universal decimal place value system for naming and calculating. Approaches to instruction and teacher education are affected by these incompatibilities, as this chapter discusses.

Historical evolution can provide a deeper understanding of the past and present in science as a means of consolidating and clarifying foundations (e.g. Jankvist 2009). We begin with a historical survey of the numeration knowledge development of pre-numeral systems and the conceptual development of numeral systems. We then track the foundational epistemological and pedagogical insights from history. Section highlights the differences between cultural practices, especially -lan guage, and their links with the universal decimal features of WNA. Post-colonial tensions, where the inconsistency between spoken and written numbers and the incompatibility between numeration and calculation appear, are also explored. Section discusses the influence of multiple communities within societies throughout history when attempts at changes are made. Different stakeholder groups in a given society may hold different goals for WNA and thus create different expectations and support within the society. Various examples are given and key comparisons are made, especially for understanding how and why curricula change.

In parallel with tendencies towards teaching mathematics in globalised ways, a shared awareness has recently evolved among teachers and researchers about the nature of mathematics through the study of its history, traditions and culture. By 'culture', we mean a set of meanings that have been historically constructed, socially transmitted and continually modified and that are embodied in our symbols and language (e.g. Barton 2008). Through this set of meanings, people communicate, perpetuate and develop their knowledge and understanding of life. History and culture shape not only number names and concepts but also the use of numbers in measurement and operations. Different languages have their own syntax and semantics, which emphasise different aspects of numbers; these may foster or hinder a deep understanding of number concepts, especially ideas about base ten, place value and operations. While a purpose of education is to support the continuity of the structures and functions that are unique to a culture and to maintain cultural identity (e.g. Leung et al. 2006), local cultures need to link to universal cultures to avoid isolation in global development. A critical issue, then, is how a cultural system reflects on its own history, language and culture, identifies the disadvantages and advantages of its system, and overcomes its disadvantages and promotes its advantages. What lessons do we learn from these reflections and from the interventions that are based on them?

Foundational Ideas that Stem from History

Introduction: The Hindu-Arabic Numeral System

According to some historians, the story of the Hindu-Arabic system (e.g. Lam and Ang 2004) is derived from the Chinese story. See (Chemla 1998) for a different historical perspective and more details in this volume, Chap. This system and its use, which was systematically presented in the Sunzi Suanjing, was transmitted through India during the fifth to ninth centuries, to the Arab Empire in the tenth century and then to Europe in the thirteenth century via the Silk Road (see Guo 2010). Mathematics historians have debated the origins of the Hindu-Arabic numeral system for years. For example, French mathematics historian Georges Ifrah (2000) argued that as the Brahmi notation of the first nine whole numbers was autochthonous and free of any outside influence, the decimal place value system must have originated in India and was the product of Indian civilisation alone. In contrast, Lam and Ang (2004) argued that there is no early Indian text or evidence to show that it was used earlier there than in China. Early texts and evidence show that the Chinese used the rod numeral system continuously for almost 2000 years. This historical fact is not well known in either the Western or Eastern communities of math education because of the limited dissemination of the conceptualisation history of place value of the Hindu-Arabic system.

According to Lam and Ang (2004), in Western Europe, before the advent of the Hindu-Arabic numeral system, few mathematicians would have been able to perform multiplication. In contrast, in ancient China, the operation of multiplication would have been commonly known as far back as the Warring States period (475–221 BCE), not only among mathematicians but also among officials, astronomers, traders and others. This could be because the Chinese rod number system used the concept of place value. It is not surprising that the Nine Chapters on the Mathematical Art includes common fractions, areas, the rule of three, least common multiple, extraction of square and cube roots, volumes, proportion and inverse proportion, relative distance and relative speed, surplus and deficit, rule of false position, the matrix notation, negative numbers, simultaneous linear equations and right-angle triangles because it was grounded in the advanced decimal place value system (Chemla, 2007).

A culture's arithmetic development may be confined or promoted by the numeral system used. For example, multiplication with large numbers could not be well supported by a simple tally system. The Hindu-Arabic numeral system is much more complex than others: it includes a principle for naming numbers, which is ten based with multiple units, and the additive and multiplicative relationships are embedded implicitly, with only the digits recorded. It is universally used in the world because every number, however large, can be easily represented and computations can be easily realised.

According to the Discussion Document (this volume, Appendix 2), historical reconstruction was in the foreground of WNA. For a better understanding of this system, a brief conceptual development of the numeral system[1] is examined, and associated epistemological and pedagogical analyses are carried out below. This study has two motivations: to understand the foundations of established whole number arithmetic (product) by studying the historical origins (process) and to provide insights for modern teaching by investigating epistemological obstacles (this volume Chap., esp. Sect.). Jankvist (2009) argued that 'history can not only help to identify these obstacles, it can also help to overcome them: an epistemologi-cal reflection on the development of ideas in the history can enrich didactical analy-sis by providing essential clues which may specify the nature of the knowledge to be taught, and explore different ways of access to that knowledge' (p. 237).

We examine the foundational idea of number representation development and do not describe all of the historical facts. The conceptual development of numeral systems can be classified into four types based on conceptual development progress: the tally system, additive system, multiplicative-additive system and decimal place value system. In each case, different strategies to realise operations were invented; these are explained below. Progress could better mirror the development of number structuring and some epistemological obstacles in history/learning (e.g. Jankvist, 2009).

---

[1] A numeral system (or system of numeration) is a writing system for expressing numbers – that is, mathematical notation for representing numbers of a given set using digits or other symbols in a consistent manner.

Knowledge of Pre-numeral Systems

Early Numeration Practices

Many anthropologists (e.g. Ifrah 2000; Menninger 1969) have found that some ancient cultures did not develop numbers at all. Some had names only for one and two and some up to three or four. Larger numbers were described as 'many'. In many ancient languages, words for 'two' or 'three' exist between singular and plural as a means to distinguish one from many, which is the beginning stage of the development of numeration.

One-to-one correspondence between an organised list of words – that is, the list of number names – and the units of a collection is typically considered an elementary process in counting, and it is the most fundamental stage. In many cultures (e.g. Menninger 1969), parts of the human body have been used to make one-to-one correspondence, often starting with fingers. Despite having limited sets of number names, some cultures developed quantitative practices that go beyond the greatest number available, such as the use of tally systems, and partitioned large quantities into smaller countable quantities (Baxter 1989). A society's early numerical practices are embedded in its development of a tally system, which is built on ordinal numbers, cardinal numbers and the counting principle – namely, one-to-one correspondence (Seidenberg 1962). Beyond tally numeral systems, various cultures developed different numeration systems, yet all had in common a symbol for one, the unit of ones and other symbols for collections of that unit (May 1973).

The Invention of the Counting Principle

Knowing whether quantities have increased or decreased was likely a key problem for many ancient tribal peoples. To recognise more or fewer, one of the earliest methods directly stimulated the invention of one-to-one correspondence with an intermediate collection of stones (Dorier 2015). Before representing and naming numbers, people developed several ways to evaluate quantities beyond rough estimation. For example, we can imagine that shepherds were concerned about the possibility of losing sheep when they returned from the fields at night. We can only speculate how the use of stones became tokens in one-to-one correspondence, but there is clear evidence that tokens were used in one-to-one correspondence by the use of 'envelopes holding counters to represent sets' of numbers (Schmandt-Besserat 1992, p. 190). Illustrations of artefacts were used to form records of number of tokens as representations. Several other artefacts, some dating to Palaeolithic times (15,000 BCE), such as notches and bones, are indicators of human activity related to the building of corresponding collections with specific cardinal numbers to record quantities. This may have been the beginning of the invention of the counting principle.

The Pre-structures of Number Naming

Because of rhythmic demands of oral pronunciation, no language represents numbers by articulating the same sound more than two times. That is, there are no known examples such as 'one one one one' for four or 'three three three' for nine. Words are sometimes repeated but not more than once (Cauty 1984; Guitel 1975). This implies that number names are not based on the principles of a tally system (a numeration system of keeping a record of quantities and amounts by using single strokes to represent the objects being counted). If number names exist in a given language, the list of the names of the smallest numbers is a sequence of words that is more or less long, and the words are more or less independent. Conversely, if a given language presents num-ber names for large numbers, sequences of number names beyond a threshold always have a multiplicative structure. The threshold is almost always under 100 (Cauty 1984; Crump 1990; Menninger 1969). In Chinese, and other languages such as Chunka (González and Caraballo 2015), the threshold is ten.

We consider the sequence before the threshold, as its structure is of interest. Cauty (1984) identified several types of sequences within spoken numbers:

1. Ordinal – a list of words independent of each other.
2. Ordinal with benchmarks – as described above, it begins with a list of words and is followed by benchmarks on a scale, e.g. the Panare language has benchmarks for 5, 10, 15 and 20 and counts from 1 to 4 between the benchmarks (Cauty 1984).
3. Cardinal with addition – where a number is represented by juxtaposed number names whose sum is the given number (such as how XXIII means two 10s and three 1s in Roman numerals).
4. Cardinal with multiplication – where a number is represented as a sum of products of small numbers ('digits') times units (such as how three hundred two means three times one hundred and two in Chinese spoken and written numerals).

The Oksapmin people (Saxe 1981) use body parts to recognise numbers, moving from the right-hand fingers up to the right eye – the first finger is 1 and the eye is 13. The nose is 14, and they move symmetrically from the left eye to the left-hand fingers to count from 15 to 27. This can be considered a long ordinal list. According to Cauty (1984), the ordinal-with-benchmarks type above is often confused with the cardinal type with addition/multiplication. The difference between the two lies in the grammar, which indicates movement in relation to the benchmarks. Benchmark and additive numeration may be the beginning of the idea of a base (as in base ten). However, in some languages, the names of larger numbers may be expressed in terms of smaller numbers and arithmetic operations, which may be the beginning of the exploration of number structure, e.g. $3 = 2 + 1$ and $5 = 2 + 2 + 1$, and even multiplicative forms such as $6 = 2 \times 3$ and $18 = 3 \times 6$. However, it is rare for 2 to be expressed as $1 + 1$ (Crump 1990). In Nigeria, the Yoruba numeral system is based on 20, and other numbers may be expressed by subtraction, e.g. $35 = (20 \times 2) - 5$.

This can also be seen in Roman numerals, where, for example, CX means 110, but XC means 90.

To meet the need for counting, various cultures developed the counting principle and one-to-one correspondence and named numbers with some regularity, sometimes using addition and/or multiplication, to specify quantity. From these practices, we can learn about the conceptual development of number systems, as naming and operations (addition/multiplication) were developed at the same time. The counting principle, invented number names and written number symbols are needed for the development of a formal numeral system. However, the coordination between cardinal numbers and ordinal numbers could be considered too trivial, too easy to explicitly design in many curricula and lessons. These could be reasons for long-term learning difficulty in later number development and operations.

The Conceptual Development of Numeral Systems

Tally Systems

The Ishango bone[2] notches (about 18,000 to 20,000 BCE), ancient Chinese knots in string and Sumerian marks on a clay tablet show what appear to document quantity as an 'early stone or agricultural age' vision of numerals (Mainzer 1983/1991) in almost all ancient cultures. Ifrah (2000, p. 64) points out that tally systems could be an early vision of systematic counting numbers 'first used at least forty thousand years ago'. These could be the origins of Roman or Etruscan numerals (Ifrah 2000, pp. 191–197). Such marks clearly point to the development of ancient written number representations. Tally numeral systems are among the most primitive means of recording quantity (Hodgson and Lajoie 2015). Counting up and down might be the natural calculation of sums and differences. It is the simplest (unary) numeral system, and it plays an important role in the fundamental counting action of building one-to-one correspondence between objects and names, forming a set of reciting numbers in ascending order. This could be how the first systematic conceptualisation of numbers in a set of numerals developed. Tally systems directly reflect the fundamental idea of counting for small numbers: one-to-one correspondence. Once tallying becomes an established practice, establishing a set of standard names for small numbers might be the next step, as it affords the ideas of both cardinal numbers and ordinal numbers and allows the description of a collection of objects arranged in a particular order. The analysis above indicates that counting principles could be key to the conceptual development of numeral systems. Their absence could result in counting by rote memorisation, skipping objects, counting randomly and counting an object twice or multiple times.

[2]The arithmetic interpretation of the Ishango bone has been recently contrasted by Keller (2016), claiming that further studies and references to archaeological finds would be needed.

Additive Systems

As counting needs for large numbers increased, the difficulty of remembering many number names led people in many societies to the idea of grouping ones, using special abbreviations for repetitions of symbols and inventing a new object or symbol to signify this quantity (Groza 1968). This grouping may be the first step in coherently combining the different records to make structural organisation common across counters (Bass 2015). It denotes the mathematical abstraction of the numeration system in which the number represented by written numerals is simply the sum of the value each numeral represents. It also requires new symbols for different groups of ones and new strategies to enumerate the new collections. The different sized groups could be the beginning of the idea of multiple units. This kind of grouping – or re-counting different cardinal collections – was developed in all ancient civilisations to count large numbers (e.g. Bass 2015). Sumerians (~3500 BCE) initially used a tally system comprised of collections of small cones to represent collections of items (Schmandt-Besserat 1992). Over time, they replaced 10 small cones with a small ball, 6 small balls (or 60 small cones) with a larger cone and 10 larger cones with a cone of the same size with a round hole in its centre, thus using a mix of bases 10 and 60. These objects were packed in a spherical clay container that had to be broken to identify the inside number. Later, these objects were represented with marks on the surface. Eventually, the objects were abandoned and only their representations were used. Using cuneiform features, marks of wedges and corners – specifically, one vertical wedge for 1 and one corner for 10 – were written on a clay tablet. Dating to around 3300 BCE, this could be the first known written additive numeral system. An updated numeration system was constructed with a set of symbols, called numerals, together with a set of rules for writing to represent numbers.

Around 1500 BCE, the Egyptians invented a hieroglyphic written additive numerical system in base ten. Around the fourteenth century BCE, the main concept used in most Chinese numerals in the oracle bone script found on tortoise shells and animal bones was grouping, which partly formed an additive system. It is interesting to note the numerals for 2 and 3, are still used in daily language. Ancient Roman numerals, such as CXXXV for 1 hundred, 3 tens and five, recorded numbers using the concept of grouping ones and regrouping into higher units. Although they used different grouping approaches at each step, many cultures (Sumerian, Babylonian, Egyptian, Greek, Roman, Arabic, Chinese, Mayan, Aztec, etc.) developed or used many-levelled additive numeral systems based on the principle of successive grouping. Thus, historically, many numeral systems were developed by progressing from tally systems to additive (grouping) systems in which multi-units with additive relations (no multiplication relations) were developed. The analysis above indicates the understanding of multi-units as necessary to deal with large numbers and of the possibility of learning difficulty in counting because of the different units and unit conversions from the tally system.

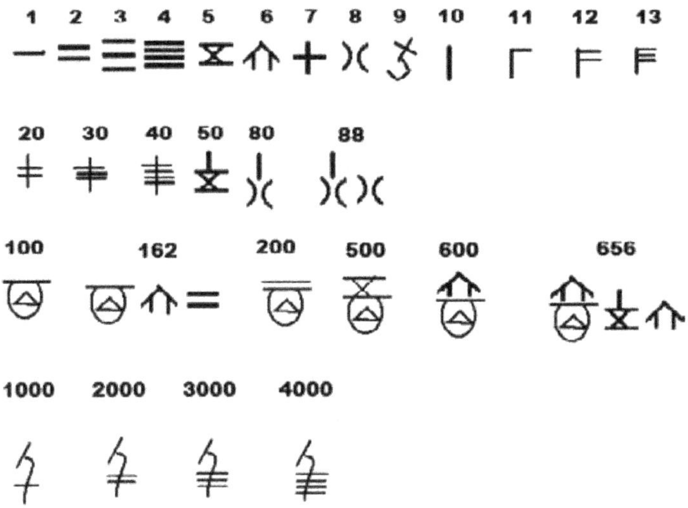

Fig. Shang oracle bone numerals from the fourteenth century BCE

Multiplicative-Additive System

The additive systems above have been rather common inventions. Many historical examples indicate that the multiplicative concept, most often with irregular forms at the beginning, has also been a common invention for representing larger numbers with a simpler approach. Counting the signs in the additive form and then naming the value of the sign lead roughly to a multiplicative-additive system. For example, the Roman numeral CCC (300) is called *trecenti* in Latin, from *tres* (three) and *centum* (hundred). It is a numeration system in which the number value should be the sum of the products of units indicating how many of each unit are considered, where the multiplicative notion is added. Here, C is a unit rather than a number in an additive system. In such systems, there is a different symbol for each power of ten and for each number from one to the base minus one.

The ancient Chinese numeral system found on bones and tortoise shells of the late Shang dynasty in the fourteenth century BCE was the first multiplicative-additive system based on the decimal system and was both additive and multiplicative in nature. Here is a selection of Shang oracle bone numerals (Martzloff 1997; Needham 1959) (Fig.).

Here, 200 is represented by the symbol for 2 and the symbol for 100, 3000 is represented by the symbol for 3 and the symbol for 1000, etc. (Fig.). The addi-tive nature of the system means that symbols were juxtaposed to indicate addition, so 4359 was represented by the symbol for 4000 followed by the symbol for 300, the symbol for 50 and the symbol for 9 (Fig.).

As this was not a positional system, there was no need for a zero (Fig.).

Guitel (1975) classified this as a hybrid system. A number of additive systems evolved into multiplicative-additive systems (Chinese, Mayan, etc.), but most

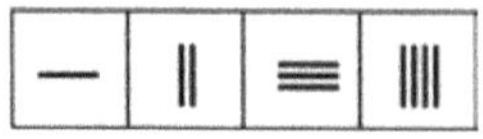

Fig. Representation of 1234 on a horizontal counting board

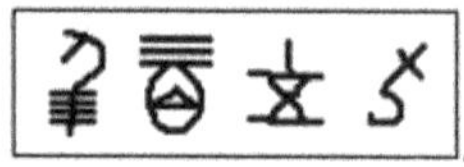

Fig. Representation of 4359

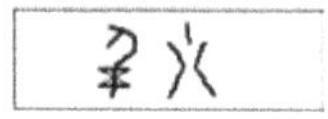

Fig. Representation of 5080

remained additive (e.g. Roman written numerals, Egyptian, Greek). Multiplicative systems improve calculation speed. However, the difference between multiplication and addition could cause concept confusion and learning difficulty, and number naming using both multiplication and addition could present an epistemological obstacle.

Decimal Place Value System

Both the decimal numeral system and positional notation or place value notation can further simplify arithmetic operations because of the use of the same symbol for different orders of magnitude (e.g. the 'ones place', 'tens place', 'hundreds place'). By using 1, 10 and 100 as numeration units (not number names in an additive numeral system) and both multiplicative and additive concepts, a much more advanced numeral system, the decimal place value system, was invented, in which a number can represent quantity with multiple decimal units. Both the digit itself and its referring numeration unit determine the value that a digit represents. The numeration unit that a symbol occupies determines the value of the unit, and the symbol itself determines how many of these units are being represented (Groza 1968). Using numeration units, calculations in the place value system are quite different from those in the tally and additive numeral systems. The numeration units and their conversions are the key to calculations. Addition should be carried out with two numbers with the same units, and numbers with different units should be converted to the same units using the following conversion rate: 1 thousand = 10 hundreds; 1 hundred = 10 tens (1 in the third place); 1 ten = 10 ones (1 in the second place); etc.

The Chinese counting rod system and the Hindu-Arabic numeral system are decimal place value systems (Japan, Korea and Thailand imported the Chinese decimal system (Lam and Ang 2004)). About the fourth century BCE (the West Zhou dynasty), the first place value system using counting rods came into use (Guo 2010; Martzloff 1997). Numbers were represented by small rods made from bamboo (Zou 2015) and used on a counting board (this volume, Chap.). A number was formed in a row with the units placed in the rightmost column, the tens in the next column to the left, the hundreds in the next column to the left, etc. (Fig.).

A zero on the counting board was simply a blank square. *Sun Zi's Suanjing* (孫子算經500 CE), the earliest extant treatise, described how to perform arithmetic operations on the counting board and gives instructions on using counting rods to multiply, divide and compute square roots. Though humans have always understood the concept of nothing or having nothing, the symbol of zero was used to represent the 'zero' first in the Hindu-Arabic numeral system. This was the first time in the world that zero was recognised as a number of its own, as both an idea and a symbol (Martzloff 1997). That place-value notation with the same ideas of zero emerging in two very different settings aimed to make more efficient systems to represent any number. Xiahou Yang's Suanjing (夏侯陽算經500 CE) explains not only positive powers of 10, but also decimal fractions as negative powers of 10 (Martzloff 1997), applying positional notation to the decimal fraction ring. Thus, decimal place-value notation emerged as a more efficient system for calculations of both whole numbers and fractions.

Modern Theoretical Approaches

Below, two schemes or 'theories' of place value numeral systems are proposed to describe written positional systems. The first is *classical theory*. It belongs to traditional arithmetic treatises, such as those of Bezout and Reynaud (1821) and Ryan (1827). This theory has been used in France to teach positional notation for centuries. The second theory belongs to academic mathematics.

The words for units used in numeration – that is, the words 'ones', 'tens', 'hundreds' and so on – are henceforth called *numeration units*. Numeration units are built one after the other in the following way. (1) The first ten numbers are built one after another, starting with the unit one and then adding one to the previous number, forming the numbers one, two, etc. (2) The set of ten ones forms a new order of units: the ten. (3) The tens are numbered like the ones were numbered before, from one ten to ten tens: one ten, two tens, etc. (4) Then the first nine numbers are added to the nine first tens: one ten, one ten and one one, one ten and two ones,… two tens, two tens and one one, and so on, forming the first 99 numbers. (5) The set of ten tens forms a new order of units: the hundred and so on. Numbers' names are presented as a literal translation built on units' names (adapted from French): 'Say the tens, then the ones'. For example, as three tens is thirty and four ones is four, then three tens and four ones is thirty-four. The rules have exceptions, however. For instance, the usual name of ten-one is eleven. Finally, after building the numbers, the positional notation is stated. To write numbers without writing the units' names, it is sufficient to juxtapose the numbers of units of each order with the ones on the right side; then, each place represents a unit that is ten times larger than the one on its right. Places that are not represented are marked with the sign 0.

The current reference knowledge for place value in academic mathematics is based on the polynomial decomposition of a whole number n in a given base r: $n = \sum a_i r^i$, $0 \leq a_i < r$, which is a much more generalised expression of classic theory with the particular formal abstractions that characterise modern mathematics. The current proof of the existence and uniqueness of the decomposition in a formal way involves Euclidian division. The positional notation is defined as the juxtaposition of the coefficients of the polynomial. This theory, which belongs to advanced algebra, is henceforth called *academic theory* (following Bezout and Reynaud 1821).

Both of the approaches presented above provide multiplicative descriptions of positional notation. The multiplicative description and the recourse to exponents are not necessary when devices for computing, such as the *suàn pán* or abacus, are used as the device itself embodies this convention.

In sum, the decimal place value system and computation were handed down to us by our ancestors and underwent improvements over time. It is worth determining which aspects were improved and the reasons for the changes. We have described several steps, each requiring concept and relation development. From the tally system to the additive system, multiunit notions – that is, the grouping of units – are critical. Grouping units simply means a counting process using larger units, which must also rebuild the multiplicative relation with lower units. Multiplication also further simplifies the repeated addition relation, which advances the abstraction process of counting. The development of this elegant positional base-ten system took place over a long time; therefore, the fact that understanding numbers is complicated should not be surprising. Without this notation, one would encounter the same difficulties that peoples of ancient cultures encountered in large number and fraction development.

Epistemological and Pedagogical Insights from History

Pedagogical Insights from the Pre-history of Numbers

The pre-history of numbers and the invention of small numbers can provide insights into the beginning of the teaching of numbers and learning by young children. One-to-one correspondence is likely an essential step towards the concept of numbers: the recognition of quantity as a property of collections. History as described in the previous section(s) shows a double role of one-to-one correspondence: the intermediate collection of objects such as stones and the intermediate collection of words, e.g. the number names. As Dorier (2000) advised, history can be used to reconstruct an epistemologically controlled genesis that takes into account the specific constraints of the teaching content. The fundamental situation of numbers conceived by El Bouazzaoui (1982) and Brousseau (1997) belongs to such a programme. It has to do with quantity (not yet numbers) and can be expressed as building the same cardinal number collection to a given collection. Typically, the task is, 'Look, there are rabbits here. Go and bring carrots so that each rabbit can eat. That is: each rabbit should have one carrot, no more, no less'. It can be observed that spontaneously, a young child does not count even if she knows a sequence of number names. From this general situation, several steps can be conceived, taking into account didactical variables (Brousseau 1997) – that is, conditions on the tasks to be achieved that can change what children learn. For example, are paper and pencil available? If so, children can draw carrots, thus making a list of what is needed, which is an intermediate collection. How is the size of the collection – e.g. [2–6], [6–12] and [12—100] – related to children's knowledge of the sequence of number names? Are the rabbits visible from where the carrots are? They can be drawn and arranged in specific ways, such as on a die or in several areas of a sheet of paper to foster concepts of, for example, three in one area, two in another and so on (Briand et al. 2004; Margolinas and Wozniak 2012).

Understanding Numerals' Uses: To Write, to Compute, to Talk

Not all systems have followed the same development. For example, China never had a recorded additive system. It is interesting to note that in the West, written Roman numerals are additive, while the corresponding spoken number names are multiplicative. For example, XXX is *trīgintā* in Latin, where *tres* is 3 and *decem* is X (ten). Spoken numerals, especially for large numbers, belong to a pre-multiplicative-additive system. However, in contrast to speaking and writing, the number concept for computation is positional.

The positional principle for the Old Babylonians (ca. 2000–1600 BCE) was in base 60. Its digits are of the written additive system in base ten as they are all under 60. In the scribal school of south Mesopotamia (ca. 2000–1800 BCE), positional notation was only used for computation, never to express measurements, which used only rather small numbers, written in additive form, associated with a developed system of units (Proust 2008, 2009). The ancient Romans and medieval Europeans did not write positional numbers but used the additive system of Roman numerals even though they computed with a positional abacus, which embodies the positional principle used by the Greeks, Old Babylonians (Høyrup 2002) and Chinese (Fernandes 2015). The position of each digit within a number denotes the multiplier (power of 10) multiplied by that digit. These tools were constructed using principles similar to the abacus: a board with columns into which identical objects are put, where all of the objects in a given column indicate the same value, which is generally that of a digit of the additive system (or of the unit if it is a numeration unit system), where adjacent columns contain objects representing two successive digits (two successive units in a numeration unit system). A key feature is that if the ratio of the digits between two adjacent columns (with the lesser on the right) is n, n objects in the right column can be replaced by one object in the adjacent left column without changing the number. If the ratio between the columns is n (which would thus be the base of the system), one simply moves the objects in the adjacent left column and multiplies by n.

The case of the Incas features incompatibilities between written numbers and calculations. The Mayans and Aztecs developed a base-20 numerical system, while most of the cultures in the Andean region developed a base-ten system. The Incan civilisation used a cord system to make alphanumeric records to code information and solve numerical problems. These *quipus* are systems of strings with different colours and knots. The position of the cords, the types of knots and the colours of the strings are elements of its logical-numerical nature (Ascher and Ascher 1981). However, the Incas used the *yupana* to make calculations using a base-ten system (González and Caraballo 2015).

When using a counting board, the successive steps of a computation in progress disappear. In the Old Babylonian case (ca. 2000–1600 BCE), as in the Chinese one, it seems that the positional notation was not used to record quantities. In the Old Babylonian tablets of Nippur (school of scribes), positional notation writing seems to have been used to indicate computational algorithms (Proust 2009) or to record the steps of a computation on a tool (Høyrup 2002; Proust 2008). In ancient Chinese books, writing with positional notation was used to explain the locations of the rods or beads on the tool and especially to explain computational algorithms (Chemla 1996; Lam and Ang 2004).

From the survey above, we come to the surprising conclusion that in many Western civilisations, to express large numbers, the written systems are additive, and the spoken systems are multiplicative, while the tools used for computations embody a positional principle. In other words, a large number written in an 'additive' way is often 'spoken' in a multiplicative way and, if used in a computation, is expressed in a 'positional' way using physical structure rather than written symbols. The invention and subsequent widespread adoption of the Hindu-Arabic system changed this, as this unique system is used to write and compute both positionally and multiplicative additively.

Understanding the Conceptual Changes in the Development of the Decimal Place Value System

Memorising the Multiplication Table

Ancient Egypt, ancient India and ancient Rome did not develop multiplication tables because of the additive systems they used. However, ancient China, Greece and Old Babylon did (Høyrup 2002; Menninger 1969). It is reasonable to imagine that the multiplication concept could be difficult in the current Western curriculum (Beckmann et al. 2015). In 1989, the National Council of Teachers of Mathematics (NCTM) developed new standards recommending reduced emphasis on the teaching of traditional methods relying on the rote memorisation of multiplication tables. NCTM made it clear that basic multiplication facts must be learned in the Common Core State Standards for Mathematics in the USA (Common Core State Standards Initiative, 2012). Historically, in China, the multiplication table has been a more stable part of the base-ten rod and abacus calculation curriculum, and there has been no controversy about memorising it (Cao et al. 2015). It is regarded as the foundation of multiplication/division column calculation and plays an important role in calculating rapidly and accurately (Cao et al. 2015). It includes the essential facts of column multiplication, which are similar to the function of number naming 1–9 for addition and subtraction. Therefore, the importance of memorising the multiplication table for multiplication/division calculation curriculum development and instruction should not be underestimated.

Unit Conversions

According to the above analysis, the one-to-one relationship of adjacent unit types in the additive system needs to change to the ten-to-one relationship of the place value system. In a simple additive system, each symbol has a fixed face value. Because in the base-ten place value system, the value that a digit represents is determined by a combination of two factors, its own face value and its reference unit within a numeral, a single symbol can represent different values depending on its referent unit. By this use of numeration units, calculations in a place value system are quite different from those in tally and additive numeral systems. The numeration units and their conversions are the key to calculations: addition should be carried out with two numbers with the same units. Numbers with different units need to be converted into the same units by the conversion rate 1 thousand = 10 hundreds, 1 hundred = 10 tens (1 in the third place), 1 ten = 10 ones (1 in the second place). Therefore, the composition and decomposition of a higher unit is a key procedure for addition/subtraction that cannot be realised by the counting approach in tally and additive numeral systems. Thanheiser (2015) designed an intervention based on a history of numeration with successive steps – tally, additive and positional base-ten systems – and then performing computations with these systems to reinforce place value for teacher education. Thanheiser argued that the unit conversions can be difficult because they do not develop naturally in calculations in the additive system: counting on, counting down and the doubles approach.

Foundational Ideas from Language and Culture

Number language is the first cultural symbol a human encounters in mathematics, and it appears more salient than other teaching and learning factors in relation to WNA. The notions of numbers are first learned as a linguistic routine of counting through which the number names are perceived as sign systems or cultural semiotic systems that enable the symbolic representation of knowledge (Goswami 2008). There are few studies on conflicts among number naming, cultural identities and universal decimal features and how local languages are linked to universal Hindu-Arabic numerals. In the following, these two questions are discussed.

Whole Number Naming: Universal vs Cultural

A clear example of such conflict is in whole number naming, where history and culture shape not only number names and conceptualisations but also their use in measurement and operations. When analysing this phenomenon, two situations emerge: the first one is what remains from ancient times, which contaminates a more recent and coherently organised cultural system at the oral level, such as the case of spoken numbers in Denmark. The second situation relates to the lively colonial process related to different uses of language fed by political and social interactions that continuously develop over the years, such as spoken and written numerals in post-colonial areas such as Algeria and Guatemala. Analysis of the two situations is possible thanks to direct reports in the working group and provides interesting insights into the historical evolution of elementary arithmetic and how it can affect number names and result in learning difficulties. Importantly, both situations may offer opportunities to enhance the learning and teaching of numbers and contribute to the awareness of a cultural identity if sensitively handled by teachers.

The Danish Case: The History of Number Names in Denmark

As presented by Ejersbo and Misfeldt (2015), in a typical situation, the effects of the past on spoken number names in Denmark have led to problems for many children. Some number labels are rooted in ancient names, reflecting a primitive non-decimal system. One hundred years ago, 'eighty' was said as 'four times twenty', but in daily speech, it became 'firs', which is close to 'fire' or 'four' in Danish. About 50 years ago, this abbreviation was made official, as 'times twenty' was dropped for counting numbers, although it was retained for ordinal numbers. 'Seventy' in Danish is said as 'half four' and previously as 'half four times twenty'. 'Half four' is actually a half taken to four (from three), so it means 'three and a half' (as in German time counting, where 'half four' means 'half past three'). Thus, as Danish practices inversion for spoken numbers between 13 and 99, 'seventy-three' is said literally as 'three and half-four'. As the old roots are unknown to most students, the names of the tens up to 100 are taught through rote learning, and the underlying rules are not addressed. This has caused difficulties for Danish students, many of whom are insecure about reading and writing two-digit numbers for their first 3 or 4 years of schooling (Ejersbo and Misfeldt 2015).

Thirty years ago, it was common practice to address these difficulties at a cultural level by offering opportunities both for teachers and for children to reflect on the roots of the mathematical notion. Examples were used at cognitive and pedagogical levels to highlight relevant features, such as the present decimal positional system (base 10 in comparison with base 20) and the additive nature of representation in a given base. Today, these opportunities are not exploited in the first years of school, as there was little evidence to support the learning of the number names (possibly due to how the teaching had been organised). Danish students continue to struggle with combining the words for a number, its cardinal value and its digit representation. Thus, while, a language changes according to how it is spoken in everyday life, it seems difficult to change number names because of cultural continuums, even though doing so would help students better understand numbers.

The Algerian Case: Language Diversity in the Post-colonial Era

Many previously colonised countries retain the influences of their old political masters, which are often reflected in their educational systems, such as in the organisation of schools, the content of the curriculum and the languages in which specific subjects are taught. Young children learn to count in their native language, and this is encouraged until they begin formal schooling. Thus, in their early years, when teachers seek to develop a strong number sense, they can find it challenging to learn WNA in a 'foreign' language. Unfamiliar words and new conceptualisations often lead to identity problems for children.

Azrou (2015) studied the situation in Algeria. Over the last 50 years, as the power and cultural relationship between France and Algeria changed, a range of decisions were made by the authorities regarding the teaching of mathematics at different levels. For example, at the university level, all mathematics is taught in French, but at the school level, while formulae are written in the Latin alphabet and read left to right, mathematical terms are written and spoken in Arabic (right to left). The difficulties young children experience are exemplified in the Algerian context. Four languages are spoken (classical Arabic, dialect, Berber and French), and oral and written words are shared differently by different social groups. For example, although numbers are taught in classical Arabic in schools, some children from particular communities speak only a dialect and/or Berber. At this stage of schooling, because of inversion, children are expected to write numbers from the lowest-value digit to the highest – that is, from right to left like word writing. This expectation changes at grade three, at which point all children are expected to engage with numbers not only in Arabic but also in French, where words are written from left to right and where there is no inversion above 16. Some children write numbers from the lowest to the highest-value digits in the direction of word writing for both languages.

The analysis of these two situations provides some insight into the language evolution of elementary arithmetic and how it can affect number naming and result in learning difficulties. Obviously, whole number language conflicts between universal, cultural and colonial naming can cause meta-level learning difficulty. Therefore, a bridge between curriculum and instruction is needed to link local languages to universal Hindu-Arabic numerals.

The Incompatibilities Between Spoken Numbers, Written Numbers and Numeration Units Within 100

When Anglophone speakers learn how to tell quantities and then write (in digits) 'what they hear', they are faced with the difficulties of two changes in the number structure: 11 and 12 sit outside the 'teen' numbers, 13–19, where an inversion structure in the spoken names (the ones-place digit, not the tens-place digit, is spoken first) is used (thirteen, fourteen, fifteen, sixteen, seventeen, eighteen and nineteen). There seems to be no logic to this disturbance to what is otherwise a strong pattern in English's number system in English. Thus, these words have to be learned and remembered. In contrast, the Chinese system provides a logical way of reading: 11, 12 and the 'teen' numbers (ten-one, ten-two, ten-three, ten-four, ten-five, ten-six, ten-seven, etc.) do not need learning or remembering in quite the same way. Another difficulty is the inversion structure (as demonstrated in the Algerian and Danish cases above) between spoken numbers (e.g. the ones are said before the tens in 14) and written numbers (e.g. the tens are written before the ones in 14), which could cause learning difficulties. The cognitive challenges of the number learning of 11–19 have long been recognised as 'the trouble with teens' (Miller and Zhu 1991). This might relate to the controversial question of place value or quantity value in the Western curriculum (e.g. at what stage should the 3 in 38 be identified as 3 tens rather than 30 (Askew and Brown 2001)?).

Houdement and Tempier (2015) identified three number systems in the process of knowing numbers: written numbers (written as 56, showing face value), spoken numbers (spoken as fifty-six, showing quantity value, number size made of ones) and numeration unit numbers (presented as 5 tens and 6 ones, showing place value, number size made of numeration units). Generally, a spoken number system is rooted in local oral language, and it naturally follows the grammar of local language and thus directly shows cultural identity. It is often irregular in Western languages (72 = *soixante-douze* in French, i.e. sixty-twelve). It is mainly learned at home as an inherited mother language. The spoken names are developed as sounds connected to the numbers of objects in the sets. Written numbers are mainly learned at school as a second language. Teaching the third system (the numeration unit numbers, showing place value) specifically and gradually in relation to the two others might facilitate understanding of (1) the base-ten place value system (Tempier 2013), (2) computation algorithms (Ma 1999) and (3) the decimal form of rational numbers. In English, numbers' names are those of the units, but not in French. In France, the names of the second, third and fourth decimal numeration units for whole numbers are not used in the everyday language (Chambris 2015).

In many languages of East Asia and elsewhere throughout the world (such as several African languages), spoken numbers are similar to numeration unit numbers. Written numbers are numeration unit numbers without the numeration units, so the relationships among them are simpler than the European ones. In the Hindu-European language area, students must memorise many spoken, written and numeration unit names without connecting them logically with the written names ('ten' is not heard between 11 and 16). A way to achieve a logical link is to translate written names into numeration unit names and vice versa. This is related to the role of numeration units in the curriculum in many countries. For example, in America, Fuson, Smith and Lo Cicero (1997) explored 'tens' and 'ones' in English (and 'decenas' and 'unidades' in Spanish), but did not refer to 'units' but to 'tens-and-ones words' and also to 'position' in Spanish. In French textbooks, numeration unit names – e.g. ones, tens and hundreds – disappeared from the curriculum in the 1970s; they may remain as place names. It is frequent not to find the relation 1 hundred = 10 tens in current French second and third grade textbooks (Chambris 2015). Tempier (2013) observed three third grade teachers and found that only one of them explicitly referred to the relations among units, although all three used the units' names to describe places. Numeration units completely disappeared as units in the 1980s, but they remained as places (position) and may have reappeared in other forms and practices since 1995.

Links and Incompatibilities Between Numeration and Calculation

It seems that hundreds and thousands play a more visible role in spoken Western languages than ones and tens because the latter do not function as units in many spoken languages within 100. Thus, an opportunity to build an understanding of the numeration structure is lost. As mentioned above, the Chinese language stresses that spoken numeration recalls both the number name and numeration unit based on the language categories of classifiers (Sun 2015), which specifies the value of digits in a clearer way. The corresponding teaching and learning of numbers and calculations emphasises how to compose/decompose a numeration unit by using a composition or decomposition approach (e.g. 1 ten = 10 ones). Compared with many Western curricula, counting and memorising number facts is emphasised, especially in the teaching and learning of 1–2 digit calculations. In Chinese, the result of 8 + 7 is said as ten-five. The number name fosters the use of the make-a-ten method: eight plus two, ten plus five as the result is given in the procedure. In the same manner, to compute 40 + 10, one has to say that four tens and (one) ten is five tens. If the language does not stress the unit of tens, children will obtain various answers. For example, in French, 8 + 7 may be 7 + 7 (double) fourteen and one. Young children memorise 'doubles' rather quickly, and then there is only one (easy) step left to find 15. To make a ten to compute 8 + 7 requires the following: (1) to make ten with 8 (take 2), (2) take 2 from 7 (5 left) and (3) transform 10 and 5 into 15. This last step does not exist in Chinese, but it does exist in languages that do not use ten to compose the names of numbers between 11 and 19. In many European languages, the cognitive cost of counting addition up to 20 with doubles (with additional counting up to 3) is probably often less than that of using the make-a-ten method. For example, Portuguese textbooks suggest 'doubles', 'doubles plus 1', 'compensation' (e.g.

6 + 8 = 7 + 7 = 14) and 'reference numbers' (6 + 7 = 5 + 1 + 5 + 2 = 10 + 3 = 13) in addition calculation and 'counting back', 'the use of tables for the addition to subtraction' and 'identifying inverse operation of subtraction as addition' in the subtraction calculation (Sun et al. 2013). In contrast to the Chinese curriculum, the make-a-ten method, a core concept of addition and subtraction within 20, is often regarded as the foundation of the place value of tens (Sun 2015). It lays the foundations for multi-digit calculation by maintaining coherence between two-digit and multiple-digit addition/subtraction, which may not be the case in most Western curricula because the tens in a teen number name do not play the role of the tens place but are a quantity value role (e.g. 10 + 3 is said as thirteen). Most European languages break away from clear regularity in respect to the base-ten place value system, especially with the numbers from 11 to 19 and 19 to 99 (Ejersbo and Misfeldt 2015). Fuson et al. (1997) found that English-speaking children perform badly with the make-a-ten method but suggested that material showing tens in teen numbers might be fruitful.

The Chinese curriculum uses the make-ten method in the first grade to develop the concept of place value and generally assigns at least 30 hr (1/2 time of the first term) for the conceptual foundation of addition/subtraction, with regrouping as its core practice. Specifically, the make-ten method has been designed critically for the composition of a tens unit, which is also an important aspect of the concept of place value (Ma 1999) and in understanding the concept of addition with regrouping inherited from the principle of the bead calculation tradition (Sun 2015). Ruthven (1998), like many Western scholars, argued that close scrutiny of the mental calculation strategies used by children for the four basic operations suggests that there is no evidence of what is normally understood by place value in their methods. Thompson (1999) claims that mental calculation strategies utilise what has been described as the quantity value aspect of place value (56 seen as 50 and 6), whereas standard written calculations necessitate an understanding of the column value aspect (56 seen as 5 tens and 6 ones). This subtle but important difference has implications for teaching, as it could provide a possible reason why Chinese teachers demonstrate greater conceptual understanding of subtraction in regrouping when decomposing a higher value unit compared with their American counterparts (Ma 1999).

How to Bridge the Incompatibility: Some Interventions

Because of the numerous difficulties in number names in many languages, interventions for bridging the incompatibilities between spoken numbers, written numbers and numeration unit numbers are critical. Numeration unit number language in many Eastern countries uses place notion, which is fundamental to the computation. What kinds of didactical language inventions can we use for WNA to better support the conceptual development of numbers and computations?

Several kinds of intervention on this issue were reported at the Macao conference:

1. Ejersbo and Misfeldt (2015) introduced a didactical linguistic invention: regular number names. These more 'logical' number names are formed with the names of the powers of ten, which are those of units in Danish, like in Chinese. This enhanced the acknowledgement of number structure/regularity and improved the rote learning of number names and computation. A similar invention (Fuson et al. 1997) 'allowed all children to enter the conversation about place-value meanings and 2-digit addition with regrouping before some of them had fully mastered the English number word sequence to 100' (Fuson 2009, p. 346).
2. Sun (2015) introduced a Chinese approach to enhance reasoning by connecting the three core concepts of addition, subtraction and numbers together in the Chinese curriculum. This approach is used in all chapters of addition and subtraction. a. Adding one to a number obtains its adjacent number. b. Subtracting one from the adjacent number gives the original number. By this approach, not only are the three concepts of addition, subtraction and number tied closely together, but connections are formed between them, and the concepts of inverse and equation are developed. This promotes not only doing and memorising but also reasoning. In contrast, in some Western curricula, the ideas of numbers, addition and subtraction are presented in three separate chapters, isolated from one another. This might influence mathematics learning attitudes at the beginning.
3. Considering the French curriculum, Houdement and Tempier (2015) proposed a complete system with a focus on numeration units with which all numbers – whole and decimal fractions – can be linked. Among other tasks, they introduced various ways of quantity counting, including one by one and ten by ten in tens.
4. Awareness of the teacher education curricula situation is poor. Azrou (2015) presented the initial step of an intervention project to use difficulties as educational resources. For example, by comparing the French wording of numbers between 11 and 16 with their Arabic or Berber counterparts, students may better realise the anomalous nature of French spoken numbers. By analysing their structure between 60 and 100 in base 20, they may engage in stimulating conversion exercises. From a teaching perspective, developing a clear understanding of these patterns of learning numbers and acknowledging their different characteristics would facilitate effective and meaningful communication between teachers and students and might thus help teachers deal with diversity in the classroom and more effectively scaffold children's learning.

Finally, Sayers and Andrews (2015) developed a simple, eight-dimensional framework through a systematic review of the literature on number sense. Teachers in the Anglophone world often refer to the importance of developing a strong number sense of WNA in elementary school to prepare learners for the adult world (McIntosh et al. 1992). However, psychologists consider number sense innate in all humans (Ivrendi 2011), indicating a gap between what a human is born with and what needs to be taught in elementary school. Sayers and Andrews (2015) developed and tested a new theoretical framework that bridges this spectrum, which they call foundational number sense, identifying key whole number concepts that require instruction during the first year of schooling.

In short, the foundational ideas that stem from languages and cultures indicate that local number naming can be used to build cultural identities. However, it also causes conflicts between local languages and universal Hindu-Arabic numerals, leading to learning difficulty. Therefore, curriculum and instruction must be bridged to link local languages to universal Hindu-Arabic numerals and calculations. The bridges between the incompatibilities of spoken numbers, written numbers and numeration unit numbers are critical to solutions. In particular, linking numeration unit numbers with spoken numbers, written numbers and the make-a-ten method within 100 may be fundamental for such interventions.

Foundational Ideas Influenced by Multiple Communities

In the previous sections, we provided insights into WNA from both historical and language perspectives. We highlighted how WNA developed from ancient times and how traditional cultural roots can influence and perhaps conflict with the process of learning and teaching mathematics today. According to Morrish (2013), contemporary education must actively seek change for rapid development of the economy, science and technology once they lose their advantage. Attempts at change, often triggered by international comparative studies of mathematics achievement, have had diverse reactions and consequences (Feniger et al. 2012). Such changes by policy makers and educators are interpreted and negotiated through a series of processes framed by a variety of principles and practices (Kanes et al. 2014; Leung 2014; Wiseman 2013).

Mathematics education is embedded in the four major contexts of economics and business, academic mathematics, science and technology and public and private stakeholders, and their influence should not be neglected. In the following, we focus on how the teaching and conceptualisation of WNA have been changed by academic mathematics, science and technology, and the public and stakeholders. The first case concerns the influence of economic and business and ancient China. The other cases concern the influence of academic mathematics, science and technology and public and private stakeholders in modern times in Israel, France and Canada, respectively. These cases seek to understand the how and why of curriculum changes, with a focus on the fundamental losses and gains. The choice of examples is related to reports made by participants at the Macao Conference.

The Influence of Economics and Business: A Case from Ancient China

Sun (2015) discusses how the early Chinese invented number names and calculation tools (counting rods and, later, the *suàn pán* or Chinese abacus), in which place value is the overarching principle that captures the spirit of WNA (Lam and Ang 2004). This advanced the development of Chinese mathematics. However, rod calculation speed is slow and inconvenient for calculation with large numbers, which is often necessary to calculate labour, capital and products in the markets. To meet the need for efficiency for economics and business development, the Chinese abacus replaced the ancient counting rods, improving calculation speed and efficacy (Sun 2015). The replacement of rods caused a significant deterioration in calculation rationale from the time of rod calculation (Lam and Ang 2004), because the step-by-step procedures using rods were replaced by the 'calculation songs' of bead calculation.

In *Tongwen Suanzhi* (同文算指 literally meaning 'rules of arithmetic common to cultures'), compiled by the official scholar Li Zhi-zao (李之藻1565–1630) of the Ming court in collaboration with the Italian Jesuit Matteo Ricci (利瑪竇1552–1610), written calculation that had been in common practice in Europe since the sixteenth century was introduced into the Chinese system. Compared with the methods of traditional bead calculations, the main advantage of written calculation lies in keeping a record of the intermediate steps, enabling easy checking afterwards. It also allows viewing of the procedure, facilitating the understanding of the underlying reasoning without having to memorise what is going on during the calculation. This is difficult to attain in calculation using counting rods or an abacus. Siu (2015) argued that the rationale for learning written calculation, at least once in a person's lifetime, seems to be the acquisition of understanding of the underlying principle of the basic operations in arithmetic, which is essential in future learning. He further warned that using electronic calculators in current primary schools is similar to bead calculation, where the intermediate calculation rationale is hidden. 'Ironically we would be turning back the wheel of history in some sense in that we erase the intermediate steps if we depend on an electronic calculator too much'(p. 137). To meet the growing need for a quick analytical and quantitative approach to problem solving, changes in calculation tools were needed. However, the associated calculation rationale was becoming weaker and thus needed to be addressed.

The Influence of Academic Mathematics: A Case from the Mathematics Community in Israel

Considering the depth of their mathematical understanding, it is natural to assume that mathematicians should have a role in the professional development of elementary school teachers. However, mathematicians have little experience of teaching

WNA. Furthermore, the discourse of university mathematics and its teaching is quite different from its elementary school counterpart. What might happen if members of these communities were to meet and interact? This question was investigated in a professional development (PD) course for in-service elementary school teachers in Israel that was conceived and taught by a professor of mathematics.

Cooper (2015) analysed mathematical and pedagogical discourse as related to a particular mathematical topic – division with remainder (DWR). In school mathematics, DWR is transient: it becomes redundant once students are familiar with the field of rational numbers. In advanced mathematics, DWR is generalised to Euclidian domains. Remainders of DWR are also important as representatives of the ring Z/nZ. In view of such differences, Cooper asks, how can the meeting of a mathematician and in-service elementary school teachers foster mutual professional growth?

The instructor (Rick) was struck by the problematic aspect of standard notation. The = sign denotes equality, which should be transitive. If we write 25:3 = 8(1) and 41:5 = 8(1), transitivity of equality demands that 25:3 = 41:5, which Rick considered 'complete nonsense'. His proposed solution was to change the notation:

Standard notation in Israel: 25:3 = 8(1). Proposed notation: 25:3 = 8(1:3)

The new notation is read 'eight with remainder 1 which needs to be divided by 3'. To justify the new notation, Rick claimed that 8(1) has no meaning as a quantity when the divisor is not known. However, some teachers challenged this, claiming that the 8 and the 1 are 'quantities' and that the equality 25:3 = 41:5 is an equivalence. Indeed, there is nothing intrinsically wrong with this equivalence relationship, where 8(1) represents a class of DWR exercises that have the same result (quotient and remainder). Thus, the equality is 'complete nonsense' only because it is not consistent with the mathematical horizon of fractions, where $\frac{25}{3}=\frac{41}{5}$ is indeed incorrect. Although they did not agree with Rick regarding the standard notation's deficiencies, the teachers did endorse the new notation for its pedagogical affordance in providing a smooth transition from whole numbers to fractions. Rick emphasised this affordance in presenting the remainder as having the *potential* to be divided, a potential that can be realised in fair sharing situations in which the unit can be split and will be mathematically realised in fraction arithmetic. Pedagogical affordances were further explored. One teacher attending to procedures at the pedagogical horizon appreciated how the new notation might offer a smooth transition to decimal long division, where students often neglect to divide the remainder. She felt that the new notation, in signalling an unfinished division, would help overcome this difficulty. This episode, one of many described by Cooper (2015), shows how two communities with conflicting perspectives on DWR and its notation jointly explored the mathematical and pedagogical aspects of mathematical notation and worked together to gain a deeper understanding of this surprisingly complex topic, gaining insight that was new to all parties involved. In this way, the community of academic mathematics contributed to the professional development of elementary school teachers while at the same time deepening their own understanding of school mathematics and its teaching.

The Influence of Science and Technology: A Case from the New Math Reform in France

Science and technology also affect WNA. After the Sputnik crisis, to boost science education, technological development and mathematical skill in the population, New Math was introduced in many countries. Chambris (2015) reported on the New Math reform in France. This international phenomenon (ICMI 2008; Kilpatrick 2012) affected all levels of mathematics teaching in the 1960s and the early 1970s and has had lingering effects. It had two major focuses: (1) teaching 'new' math, including renewing of the mathematical fundamentals of teaching (e.g. Griesel 2007) and (2) taking into account psychological features related to learning and child development. Two famous subjects were introduced in WNA: set theory, an aspect of contemporary math, and numeration bases other than ten (hereafter called bases) (to teach base-ten principles), an aspect of psychology (Kilpatrick 2012; Bruner 1966). This phenomenon may be interpreted in terms of the construct of *didactical transposition* (Chevallard 1985) – the transformation and adaptation of knowledge produced by the scientific community to make it suitable for use as learning objects. This transformation occurred at the end of the 1970s.

In France, from 1900 to 1960, the classical theory (history section above) was adapted in close terms in textbooks. Tasks such as 'Write in figures: 3 H 4 T 5 O' and 'convert 3 hundreds into tens' – both using the symbolic register of the numeration units – were basic and current. They disappeared as the 'bases' appeared. Students interpreted the positional notation in 'bases' as a procedure, grouping and ungrouping, and they struggled when they had no manipulatives left (ERMEL 1978; Perret 1985). The process is as follows: $\sum r_i a^i$ becomes $\sum r_i 10^i$; then, a x 1000 + b x 100 + c x 10 + d or a000 + b00 + c0 + d. Such 'writings' as 40 + 7 + 50 + 43 + 25 also appeared. 'The key issue is to familiarise children with a direct work with the writings' (our translation; ERMEL 1978, p. 17): 1, 10, 100, 1000, etc. played an increasing role. Within a few years, 'write in figures: 3 H 4 T 5 O' was replaced by 'compute 3 x 100 + 4 x 10 + 5' or 'compute 300 + 40 + 5'. Conversions were not replaced. A new symbolic register – that of the powers of ten written in figures – had emerged, and the number 1 became the only unit to be taught.

Within the transposed academic theory, the technique to obtain positional notation from 3 x 100 + 4 x 10 + 5 x 1 is to 'juxtapose 3, 4, 5'. The present implicit rules to obtain it are to (1) multiply by 100 (resp. 10) (write two (resp. one) zeroes on the right) and (2) add numbers (use a 'column algorithm'). That is, put them one under the other, aligned from the right side. In 1995, decompositions with numeration units began to come back. The way to achieve them is 3 H = 300 (due to the hundreds place), 4 T = 40, 5 O = 5; then, compute 300 + 40 + 5, that is, 345. Thus, in the contemporary period, there is a kind of hybridisation by the units of the transposed academic theory. However, numeration units indicate only the digits' places and their units, and it is common not to find the relation 1 hundred = 10 tens in present second and third grade textbooks. This provides three interpretations of positional notation: it sums up an additive relation in the latter case, the numbers of units in the classical theory, and serves as a polynomial algebraic relation within the transposed academic theory.

The influence of academic mathematics is apparent within the use of the academic theory. As in the previous case, two kinds of tension between school and academic mathematics arise: (1) different needs (units vs advanced algebra) and (2) different perspectives on a shared sign symbolism (positional notation).

The Influence of Public and Private Stakeholders: A Case from Current Curriculum Reform in Canada

Public and private stakeholders also affect WNA. The implementation of international achievement tests such as TIMSS in 1995 and OECD's PISA in 2000 prompted widespread curriculum reform, which often focused extensively on WNA. Multiple stakeholders, including educators, school personnel, business leaders and parents, often expressed conflicting arithmetic goals and pedagogical expectations (Brown and Clarke 2013). These conflicts, called the 'Math Wars', occurred in North America (e.g. Klein 2007), Europe (e.g. Prenzel et al. 2015) and China (Zhao 2005).

Although Canada performs well in international tests, the public and private stakeholders of current curricula have sparked public debate. For example, students are expected to develop flexible and mental mathematics strategies through the use of compensation methods (e.g. $54 - 37 = 54 - 40 + 3$) and properties of numbers and operations (e.g. $8 \times 6 = 8 \times 3 \times 2$) to compute. Through an analysis of online responses to newspaper articles reporting Canada's faltering PISA scores, McGarvey and McFeetors (2015) sought to understand public perspectives. The following quotation is an example of an online comment posted in response to a national news article written by mathematician opposing today's elementary curriculum:

> I learned things by rote in grade school and then later on in high school I learned how to do abstract problem solving in topics like algebra. From what I understand, the system is unnecessarily complicating kids' minds by saying they shouldn't memorize basic multiplication or learn how to do long division or carry numbers. I don't see the usefulness in that. (McDonald 2013)

Keywords in this comment such as 'by rote', 'memorise', 'long division' and 'carry' tend to trigger negative reactions from mathematics educators in much the same way that 'mathematics for understanding', 'strategy-based learning' and 'student-centred approaches' often trigger negative reactions from the public.

Rather than dichotomising the arguments, the analysis revealed two sets of shared goals: (1) students need the opportunity to reach expected mathematics goals (e.g. develop computational skills and problem-solving), and (2) essential supports must be in place for students to reach the goals of mathematics learning (e.g. knowledgeable teachers, clearly written teaching resources). Reframing criticism into mutual concerns offers a starting place for conversation that empowers communities to see commonalities in their perspectives of children's mathematics learning. In short, public and private stakeholders in Canada and in other countries have affected WNA by changing policies and goals.

Foundational Ideas Summary: Understanding the Unpredictable Long-Term Effects of Change

The four reports above describe the influence of economics and business, academic mathematics, science and technology, the public and stakeholders and the various changes in the conception of WNA and its teaching. For example, in the wake of New Math, the incorporation of academic mathematics as foundational knowledge into school mathematics may lead to fundamental loss of various elements, including numeration units. In contrast, Cooper's example demonstrates how the mathematicians' perspective must undergo didactic transposition to fit the teaching and learning of WNA. Thus, to foster change, school mathematics and academic mathematics should be combined. The professional development described by Cooper can be seen as an important step in the enactment of such a transposition. However, taking the mathematicians' perspective into account in WNA curriculum design is a complex matter. While these examples may not be applicable outside their specific contexts, they may lead us to reflect upon ways in which new goals for WNA might be defined and implemented. However, a number of themes emerge when considering the four cases as a whole related to issues that are likely to arise when defining, implementing and communicating new goals for WNA.

Deep changes seem to occur slowly, perhaps because some features of prior practices persist, but once change has begun, it seems impossible to stop. This is evident in the lingering effects of New Math long after its apparent failure. The arguments used to introduce bases in the New Math curriculum came from psychology. Yet after the turbulent period of reform, during the so-called counter-reform, the "new" mathematics – here represented by an academic theory that is, in fact, not so new – became a mathematical foundation for the teaching of place value. The effects of the changes were probably neither anticipated nor mastered. This also seems to be the case in the changes in China, especially the replacement of rods with beads. When change is implemented, awareness by educators and policy makers is needed because where there are gains, there can also be fundamental losses to elements once taken for granted. This is important because key aspects of a complex system may be almost invisible to observers and may be lost when changes are made.

New Math was partly directed by economic issues linked with the growing needs of engineers and scientists in Western societies. PISA studies from the Organisation for Economic Co-operation and Development aim to determine the extent to which 15-year-old students are prepared to face the demands of the society they will live in. Does society need globalised mathematics education and, if so, how will it manifest itself in the future? Are the PISA tests the *Tongwen Suanzhi* (above) of modern times? Where will the changes develop in relation to the economic needs of society (Siu 2015)? One might suppose that history repeats itself in the redesigning of a WNA curriculum, as it must take into account the evolution of economic societal needs, thus causing tensions such as those described above.

In the examples described above, one or more of the operating communities engaged in a genuine attempt to understand the other's perspective and to promote productive communication. Despite decades of efforts to explain the benefits of current approaches in whole number arithmetic, the arguments have not been convincing for many stakeholders. Seeking ways to identify common goals and to address unfamiliarity with current approaches is an essential step in re-engaging parents and the community in children's learning. A certain failure of New Math reform is acknowledged (Kilpatrick 2012), despite some cooperation between mathematicians, psychologists, math educators and teachers. A lack of teacher education, a lack of resources and incompatibility between academic math and school math are often cited as reasons for this failure. Finally, recurring themes arise as means to achieve changes: bridging communities to share goals and means between stakeholders and the adaptation of teacher education and resources to the new goals.

The What and Why of WNA: Towards a Cognitive Dimension

The historical, cultural and linguistic foundations of WNA set out in this chapter have influenced how an educational system develops its WNA. How students develop fundamental ideas about WNA and what is needed by teachers to nurture such ideas remain themes in the following chapters. However, several papers in this group have examined some of the different representations of how teachers might present WNA at different stages of children's development. The discussions in these papers have provided important insights into the processes that can be evoked in different communities from different perspectives. Other influential aspects, such as humans' innate cognitive abilities, are detailed in Chap., while language, artefacts and tasks used by teachers are instantiated in Chap. The delicate question of teacher education is addressed in Chap.

References

Ascher, M., & Ascher, R. (1981). Ethnomathematics. In A. B. Powell & M. Frankenstein (Eds.), *Ethnomathematics: Challenging Eurocentrism in mathematics* (pp. 25–50). Albany: State University of New York Press/SUNY Press.

Askew, M., & Brown, M. (2001). *Teaching and learning primary numeracy: Policy, practice and effectiveness*. Nottingham: British Educational Research Association.

Barton, B. (2008). *The language of mathematics: Telling mathematical tales*. New York/Berlin: Springer.

Baxter, W. T. (1989). Early accounting: The tally and checkerboard. *Accounting Historians Journal, 16*(2), 43–83.

Bezout, E., & Reynaud, A.-A.-L. (1821) *Traité d'arithmétique à l'usage de la marine et de l'artillerie, 9e édition*.

Briand, J., Loubet, M., & Salin, M.-H. (2004). *Apprentissages mathématiques en maternelle*. Paris: Hatier.

Brousseau, G. (1997). *The theory of didactical situations in mathematics (Didactique des mathematiques, 1970–1990)* (N. Balacheff, M. Cooper, R. Sutherland, & V. Warfield, Eds. and Trans.). Dordrecht: Kluwer Academic Publishers.

Brown, T., & Clarke, D. (2013). Institutional contexts for research in mathematics education. In M. A. Clements, A. J. Bishop, C. Keitel, J. Kilpatrick, & F. K. S. Leung (Eds.), *Third international handbook of mathematics education* (pp. 459–484). New York: Springer.

Bruner, J. S. (1966). *Toward a theory of instruction*. Cambridge, MA: Harvard University Press.

Cauty, A. (1984). Taxinomie, syntaxe et économie des numérations parlées. *Amerindia, 9*, 111–146.

Chemla, K. (1996). Positions et changements en mathématiques à partir de textes chinois des dynasties Han à Song-Yuan. Quelques remarques. *Extrême-Orient, Extrême-Occident, 18*, 115–147.

Chemla, K. (1998). History of mathematics in China: A factor in world history and a source for new questions. *Documenta Mathematica, extra volume ICM 1998, III*, 789–798.

Chemla, K. (2007). Observing mathematical practices as a key to mining our sources and conducting conceptual history: Division in ancient China as a case study. In L. Soler et al. (Eds.), *Science after the practice turn in the philosophy, history, and social studies of science* (Vol. 14, pp. 238–276). London: Routledge.

Chevallard, Y. (1985). *La transposition didactique*. Grenoble: La Pensée sauvage.

Crump, T. (1990). *The anthropology of numbers*. Cambridge: University Press.

Dorier, J.-L. (2000). Use of history in a research work on the teaching of linear algebra. In V. Katz (Ed.), *Using history to teach mathematics – An international perspective, MAA notes* (Vol. 51, pp. 99–110). Washington, DC: The Mathematical Association of America.

El Bouazzaoui, H. (1982). *Étude de situations scolaires des premiers enseignements du nombre et de la numération. Relations entre divers caractères de ses situations et le sens, la compréhension de l'apprentissage de ces notions*. Doctoral dissertation. Bordeaux: Université de Bordeaux 1.

ERMEL. (1978). *Apprentissages mathématiques à l'école élémentaire. Cycle élémentaire. Tome 2*. Paris: SERMAP-Hatier.

Feniger, Y., Livneh, I., & Yogev, A. (2012). Globalisation and the politics of international tests: The case of Israel. *Comparative Education, 48*(3), 323–335.

Fernandes, L. (2015, January 11). *The Abacus: A brief history* [Blog post]. Retrieved from http://www.ee.ryerson.ca/~elf/abacus/history.html

Fuson, K. C. (2009). Avoiding misinterpretations of Piaget and Vygotsky: Mathematical teaching without learning, learning without teaching, or helpful learning-path teaching? *Cognitive Development, 24*(4), 343–361.

Fuson, K. C., Smith, S. T., & Lo Cicero, A. (1997). Supporting Latino first graders' ten-structured thinking in urban classrooms. *Journal for Research in Mathematics Education, 28*(6), 738–766.

Goswami, U. (2008). *Cognitive development. The learning brain*. Hove/New York: Psychology Press.

Griesel, H. (2007). Reform of the construction of the number system with reference to Gottlob Frege. *ZDM: Mathematics Education, 39*(1–2), 31–38.

Groza, V. S. (1968). *A survey of mathematics: Elementary concepts and their historical development*. New York: Holt, Rinehart and Winston.

Guitel, G. (1975). *Histoire comparée des numérations écrites*. Paris: Flammarion.

Guo, S. (2010). *Chinese history of science and technology*. Beijing: Science Press. [In Chinese].

Høyrup, J. (2002). A note on old Babylonian computational techniques. *Historia Mathematica, 29*(2), 193–198.

ICMI. (2008). *The first century of the International Commission on Mathematical Instruction. History of ICMI* by Furinghetti, F., & Giacardi, L. http://www.icmihistory.unito.it

Ifrah, G. (2000). *The universal history of numbers: From prehistory to the invention of the computer*. New York: Wiley.

Ivrendi, A. (2011). Influence of self-regulation on the development of children's number sense. *Early Childhood Education Journal, 39*(4), 239–247.

Jankvist, U. T. (2009). A categorization of the "whys" and "hows" of using history in mathematics education. *Educational Studies in Mathematics, 71*(3), 235–261.

Kanes, C., Morgan, C., & Tsatsaroni, A. (2014). The PISA mathematics regime: Knowledge structures and practices of the self. *Educational Studies in Mathematics, 87*(2), 145–165.

Keller, O. (2016). *l'invention du nombre*. Paris: Garnier.

Kilpatrick, J. (2012). The new math as an international phenomenon. *ZDM: Mathematics Education, 44*(4), 563–571.

Klein, D. (2007). A quarter century of US "math wars" and political partisanship. *BSHM Bulletin, 22*, 22–33.

Lam, L. Y., & Ang, T. S. (2004). *Fleeting footsteps: Tracing the conception of arithmetic and algebra in ancient China*. Singapore: World Scientific.

Leung, F. K. (2014). What can and should we learn from international studies of mathematics achievement? *Mathematics Education Research Journal, 26*(3), 579–605.

Leung, F. K., Graf, K. D., & Lopez-Real, F. J. (Eds.). (2006). *Mathematics education in different cultural traditions: A comparative study of East Asia and the West: The 13th ICMI study*. New York: Springer.

Ma, L. (1999). *Knowing and teaching elementary mathematics: Teachers' understanding of fundamental mathematics in China and the United States*. Mahwah: Lawrence Erlbaum Associates.

Mainzer, K. (1991). Natural numbers, integers and rational numbers. In H.-D. Ebbinghaus et al., *Numbers* (with an introduction by K. Lamotke; H.L.S. Orde, Trans.; J.H. Ewing, Ed.) (pp. 9–26). New York: Springer. (Original work published 1983)

Margolinas, C., & Wozniak, F. (2012). *Le nombre à l'école maternelle: Une approche didactique*. Bruxelles: De Boeck.

Martzloff, J.-C. (1997). *A history of Chinese mathematics*. Berlin-Heidelberg: Springer.

May, K. O. (1973). *Bibliography and research manual of the history of mathematics*. Toronto: University of Toronto Press.

McDonald, M. (2013, September 13). Frustrated professors convince elementary schools to step back from "new math" and go "back to basics". *The National Post*. Retrieved from http://news.nationalpost.com/2013/09/13/frustrated-professors-convince-schools-to-step-back-from-new-math-and-go-back-to-basics/

McIntosh, A., Reys, B., & Reys, R. (1992). A proposed framework for examining basic number sense. *For the Learning of Mathematics, 12*(3), 2–8.

Menninger, K. (1969). *Number words and number symbols: A cultural history of numbers*. Cambridge, MA: The MIT Press. (Translated from the German edition of 1958).

Miller, K. F., & Zhu, J. (1991). The trouble with teens: Accessing the structure of number names. *Journal of Memory and Language, 30*(1), 48–68.

Morrish, I. (2013). *Aspects of educational change (RLE Edu D)*. London: Routledge.

Needham, J. (1959). *Science and civilisation in China. Vol. 3: Mathematics and the sciences of the heavens and the earth*. Cambridge: Cambridge University Press.

Perret, J. F. (1985). *Comprendre l'écriture des nombres*. Bern: P. Lang.
Prenzel, M., Blum, W., & Klieme, E. (2015). The impact of PISA on mathematics teaching and learning in Germany. In K. Stacey & R. Turner (Eds.), *Assessing mathematical literacy: The PISA experience* (pp. 239–248). Cham: Springer International.
Proust, C. (2008). Quantifier et calculer: Usages des nombres à Nippur. *Revue d'histoire des mathématiques, 14*(2), 143–209.
Proust, C. (2009). Numerical and metrological graphemes: From cuneiform to transliteration. *Cuneiform Digital Library Journal, 1*.
Ruthven, K. (1998). The use of mental, written and calculator strategies of numerical computation by upper primary pupils within a 'the calculator-aware' number curriculum. *British Educational Research Journal, 24*(1), 21–42.
Ryan, J. (1827). *An elementary treatise on arithmetic, in theory and practice*. New York: Collins & Hannay.
Saxe, G. B. (1981). Body parts as numerals: A developmental analysis of numeration among the Oksapmin in Papua New Guinea. *Child Development, 52*(1), 306–316.
Schmandt-Besserat, D. (1992). *Before writing* (Vol. 2). Austin: University of Texas Press.
Seidenberg, A. (1962). The ritual origin of counting. *Archive for History of Exact Sciences, 2*(1), 1–40.
Sun, X., Neto, T. B., & Ordóñez, L. E. (2013). Different features of task design associated with goals and pedagogies in Chinese and Portuguese textbooks: The case of addition and subtraction. In C. Margolinas (Ed.), *Task design in mathematics education. Proceedings of ICMI study 22, July 2013, Oxford, United Kingdom* (pp. 409–418). Retrieved from https://hal.archives-ouvertes.fr/hal-00834054v3
Tempier, F. (2013). *La numération décimale de position à l'école primaire: une ingénierie didactique pour le développement d'une ressource*. Doctoral dissertation. Paris: Université de Paris Diderot – Paris 7.
Thompson, I. (1999). Implications of research on mental calculation for the teaching of place value. *Curriculum, 20*(3), 185–191.
Wiseman, A. W. (2013). Policy responses to PISA in comparative perspective. In H.-D. Meyer & A. Benavot, *PISA, power and policy. The emergence of global educational governance* (pp. 303–322). Oxford: Symposium Books.
Zhao, Y. (2005). Increasing math and science achievement: The best and worst of the east and west. *Phi Delta Kappan, 87*(3), 219–222.

Cited papers from Sun, X., Kaur, B., & Novotna, J. (Eds.). (2015). Conference proceedings of the ICMI study 23: Primary mathematics study on whole numbers. Retrieved February 10, 2016, from www.umac.mo/fed/ICMI23/doc/Proceedings_ICMI_STUDY_23_final.pdf

Azrou, N. (2015). Spoken and written arithmetic in post-colonial country: The case of Algeria (pp. 44–51).
Bass, H. (2015). Quantities, numbers, number name, and number line (pp. 10–20).
Beckmann, S., Izsák, A., & Ölmert, I. B. (2015). From multiplication to proportional relationships (pp. 518–525).
Cao, Y. Li, X., & Zuo, H. (2015). Characteristics of multiplication teaching of whole numbers in China: The application of the nine times table (pp. 423–430).
Chambris, C. (2015). Mathematical foundations for place value throughout one century of teaching in France (pp. 52–59).
Changsri, N. (2015). First grade students' mathematical ideas of addition in the context of lesson study and open approach (pp. 60–67).

Cooper, J. (2015). Combining mathematical and educational perspectives in professional development (pp. 68–75).
Dorier, J.-L. (2015). Key issues for teaching numbers within Brousseau's theory of didactical situations (pp. 76–83).
Ejersbo, L., & Misfeldt, M. (2015). The relationship between number names and number concepts (pp. 84–91).
González, S., & Caraballo, J. (2015). Native American cultures tradition to whole number arithmetic (pp. 92–98).
Hodgson, B. R., & Lajoie, C. (2015). The preparation of teachers in arithmetic: A mathematical and didactical approach (pp. 307–314).
Houdement, C., & Tempier, F. (2015). Teaching numeration units: Why, how and limits (pp. 99–106).
McGarvey, L., & McFeetors, J. (2015). Reframing perceptions of arithmetic learning: A Canadian perspective (pp. 115–123).
Sayers, J., & Andrews, P. (2015). Foundational number sense: The basis of whole number arithmetic competence (pp. 124–131).
Siu, M. K. (2015). Pedagogical lessons from Tongwen Suanzhi (同文算指) – transmission of Busan (算 written calculation) in China (pp. 132–139).
Sun, X. (2015). Chinese core tradition to whole number arithmetic (pp. 140–148).
Thanheiser, E. (2015). Leveraging historical number systems to build an understanding of base 10 (pp. 149–156).
Zou, D. (2015). Whole number in ancient Chinese civilisation: A survey based on the system of counting-units and the expressions (pp. 157–164).

6

Neuro-cognitive, Cognitive and Developmental Approaches: Whole Number Thinking, Learning and Development

Joanne Mulligan, Lieven Verschaffel
Anna Baccaglini-Frank, Alf Coles, Peter Gould, Shengqing He, Yunpeng Ma, Jasmina Milinković, Andreas Obersteiner, Nicole Roberts, Nathalie Sinclair, Yanling Wang, Shu Xie, and Der-Ching Yang

Introduction

What Was Presented at the Conference: Overview

The participants of working group 2 presented a broad range of studies, 11 papers in total, related to whole number learning representing research groups from 11 countries as follows.

Two large cross-sectional studies focused on developmental aspects of young children's number learning provide a lens for re-examining 'traditional' features of number acquisition. van den Heuvel-Panhuizen (the Netherlands) presented a co-authored paper with Elia (Cyprus; Elia and van den Heuvel-Panhuizen 2015) on a

---

Electronic Supplementary Material The online version of this chapter (doi:10.1007/978-3-319-63555-2_7) contains supplementary material.

The first two authors wrote the chapter with the collaboration of other working group members.

J. Mulligan (✉)
Macquarie University, Sydney, NSW, Australia
e-mail: joanne.mulligan@mq.edu.au

L. Verschaffel (✉)
Katholieke Universiteit Leuven, Leuven, Belgium
e-mail: lieven.verschaffel@kuleuven.be

A. Baccaglini-Frank
Università di Pisa, Pisa, Italy

A. Coles
University of Bristol, Bristol, UK

cross-cultural study of kindergartners' number competence focused on counting, additive and multiplicative thinking. Second, Milinković (2015) examined the development of young Serbian children's initial understanding of representations of whole numbers and counting strategies in a large study of 3- to 7-year-olds. Children's invented (formal) representations such as set representation and the number line were found to be limited in their recordings.

In a South African study focused on early counting and addition, Roberts (2015) directs attention to the role of teachers by providing a framework to support teachers' interpretation of young disadvantaged learners' representations of number when engaging with whole number additive tasks.

Some papers reflected the increasing role of neuroscientific concepts and methodologies utilised in research on WNA learning and development. Sinclair and Coles (2015) drew upon neuroscientific research to highlight the significant role of symbol-to-symbol connections and the use of fingers and touch counting exemplified by the *TouchCounts* iPad app.

Gould (2015) reported aspects of a large Australian large study of children in the first years of schooling aimed at improving numeracy and literacy in disadvantaged communities. A case study exemplified how numerals were identified by relying on a mental number line by using location to retrieve number names. This raised the question addressed in the neuroscientific work of Dehaene and other papers focused on individual differences in how the brain processes numbers.

The Italian PerContare[1] project (Baccaglini-Frank 2015) built upon the collaboration between cognitive psychologists and mathematics educators, aimed at devel-

[1]The PerContare project was coordinated by Fondazione ASPHI onlus, with the support of Compagnia di San Paolo and the operative support of Fondazione per la Scuola of Compagnia di San Paolo of Torino.

P. Gould
NSW Department of Education and Communities, Sydney, NSW, Australia

S. He
Hangzhou Normal University, Hangzhou, China

Y. Ma • Y. Wang • S. Xie
Northeast Normal University, Changchun, China

J. Milinković
University of Belgrade, Belgrade, Serbia

A. Obersteiner
Technische Universität München, München, Germany

N. Roberts
University of the Witwatersrand, Johannesburg, South Africa

N. Sinclair
Simon Fraser University, Burnaby, BC, Canada

D.-C. Yang
National Chiayi University, Chiayi, Taiwan

oping teaching strategies for preventing and addressing early low achievement in arithmetic. It takes an innovative approach to the development of number sense that is grounded upon a kinaesthetic and visual-spatial approach to part-whole relationships.

Mulligan and Woolcott (2015) provided a discussion paper on the underlying nature of number. They presented a broader view of mathematics learning (including WNA) as linked to spatial interaction with the environment; the concept of connectivity across concepts and the development of underlying pattern and structural relationships are central to their approach.

One group of papers presented studies about other computational aspects of WNA such as the variation, efficiency and flexibility of representations and strategies for counting, mental arithmetic, written algorithms, computational estimation and word problems. Obersteiner and colleagues (Obersteiner et al. 2015) proposed a coherent five-level competence model for WNA in the lower grades of elementary school. In another study Verschaffel and colleagues (Verschaffel et al. 2015) compared two kinds of strategies for processing mental subtraction, namely, subtraction-by-addition. Another three different studies provided new insights into mental and written WNA strategies and errors by students in the middle elementary grades: He (2015) focused on cognitive strategies for solving addition and subtraction problems; Yang (2015) highlighted the conceptual difficulties of students' judging the reasonableness of results in whole number calculations; Ma et al. (2015) analysed students' systematic errors for three-digit multiplication and linked these errors to teaching strategies. While these studies generated rich discussion about the range of research questions that focused on computational processes linked to WNA, these studies were not considered the main focus of this chapter, which instead articulates the ICMI23 (Theme 2) position paper.

The Discussion of the Working Group

As in most other working groups, the eight 1-hour sessions were organised in two different forms. Whereas the first five sessions were devoted to the presentation and discussion of the participants' accepted papers, the last three whole-group sessions involved discussions, wherein two major themes were discussed. First, to what extent can the currently influential neuro-cognitive perspective, as elaborated in Butterworth's plenary lecture, act as an appropriate theoretical scope to think about early mathematical development (and stimulation of that development) or whether this perspective needs to be nuanced and enriched by other perspectives? The second major point of discussion addressed the potentialities and limitations of the methodologies utilised in the studies on children's whole number learning and development being presented in the working group (and in some other working groups) and, more specifically, of (a) the design of the cross-sectional, longitudinal and intervention studies aiming at understanding how children develop competencies with whole numbers, as well as (b) the tasks that are common to many studies measuring the understanding of magnitude of numbers, such as number comparison and number-line estimation tasks.

Obviously, the topics dealt with in working group 2 were related to those addressed in the other working groups and panels, as will become clear through the numerous cross-references that will be provided in this chapter. However, working group 2 tried to tackle these common topics from the two perspectives mentioned above, namely, the psychological and the methodological perspectives.

About the Chapter

The chapter focuses essentially on two key aspects of the Theme 2 discussion (as presented in the Discussion Document, see the Appendix 1 to this volume) that discussed neuro-cognitive, cognitive and developmental analyses of whole number learning. Its aim is to bring these perspectives into our discussion by acknowledging the realisations and promises of neuroscientific research while adopting a critical approach from a mathematics education perspective. The structure and content of the chapter are an outcome of our synthesis of key ideas following our discussions in working group 2. Thus, the chapter will (1) present, discuss and illustrate perspectives complementary to neuro-cognitive research and (2) discuss methodologies utilised in studies on children's whole number learning and development. There are five main sections.

The bulk of the first section (Sects.) provides an overview of the ICMI Study 23 Conference presentations and the working group 2 discussion.

The second section focuses on two neuro-cognitive perspectives: first, Butterworth's 'starter kit' is discussed in light of Butterworth's plenary paper (see Chap.) and his contribution to the working group discussions. Second, some related research on the triple-code model of Dehaene and colleagues (Dehaene et al. 2003; see also Dehaene 2011) is presented. Gould (2015) then draws upon some examples from the research of how quantities and numbers are transcoded and represented.

The third section provides an overview of related research from cognitive perspectives that informs the discussion for working group 2. Verschaffel and Mulligan develop this overview of the literature to complement the examples provided by working group 2 participants. Cross links with examples from other themes are highlighted.

The fourth sectiondescribes some pertinent examples of studies presented to working group 2 and applications of the perspectives described in Section: ordinality (Sinclair and Coles 2015), part-whole relations (Baccaglini-Frank 2015), additive relations (Roberts 2015), number competence (Elia and van den Heuvel-Panhuizen 2015) and counting and representational structures (Milinković 2015).

The fifth section discusses methodological issues common to neuro-cognitive, cognitive and developmental analyses of studies on children's WNA. Cross-sectional, longitudinal and intervention studies are discussed in terms of their appropriateness for investigating children's competence with whole number. In that section, we also discuss task design in cognitive neuroscience research pertinent to number learning.

In the sixth section, some tentative conclusions are drawn and implications for teaching and learning and further research are discussed.

Neuro-cognitive Perspectives

A 'Starter Kit' for Early Number

The components of Butterworth's 'starter kit' (Butterworth 2005) for early number learning are primarily focused on cardinal aspects of number and its importance for later mathematical development. In most cognitive neuroscientific studies, children's foundational competencies are related to children's general mathematical achievement as measured by standard school achievement tests of number and computation rather than other aspects of mathematical development such as spatial processes.

From a very young age, humans have an inherited core capacity for numerical processing. For example, the process of *subitising* refers to the immediate and accurate estimate of one to four objects without serial enumeration. Another core process is described as representing non-symbolic numerical magnitudes on a mental number line. Symbolic representations (3, 100, ½, 3.17...) are gradually mapped onto these non-symbolic representations. These magnitude representations are commonly assessed by means of subitising, magnitude comparison and number line estimation tasks. Examples are shown in Figure.

Butterworth (see Chap.) describes these two foundational 'core systems'. Deficiencies in these core systems may contribute to low numeracy. He refers to an 'object tracking system' that has a limit of three or four objects and is thought to underlie 'subitising'. Another core system is the 'analogue number system' (ANS). 'The internal representations of different numerical magnitudes can be thought of as Gaussian distributions of activation on a 'mental number line'. It is typically tested

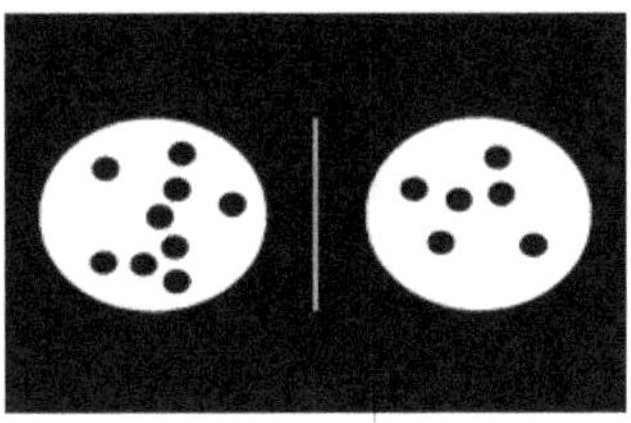

Fig. Which is the larger set?

Fig. The line begins at 0 and ends at 10. Where is the number 6 located?

by tasks involving clouds of dots (or other objects) typically too numerous to enumerate exactly in the time available' (p. 480).

Butterworth refers to the study of Australian Aboriginal children that exploited their known visual strengths to solve accurately non-standard arithmetic tasks where they had no number words to describe the quantities (Butterworth and Reeve 2008). These children matched the spatial patterns of the addend and augend sets. The findings suggested that there are various models for number that are not necessarily one-dimensional such as in the mental number line and these can be two-dimensional in nature.

Some key findings are drawn from Butterworth's research: (1) numerical magnitude understanding is positively and predictively related to (general) mathematics achievement and (2) numerical magnitude understanding can be improved by means of game-based intervention programmes, although the transfer effects from those games to mathematics learning more broadly is still rather small (see Chap.).

Neuropsychology and the Triple-Code Model

Neuropsychologists have sought to understand how brain functioning influences cognition, including mathematics and whole number learning (Dehaene 2011). Simple models associated with number processing have been proposed and tested, and the main areas of the brain identified as being activated in number processing have been refined.

Dehaene et al. (2003) proposed a triple-code model of working with number, consisting of three components: verbal, visual (numerals) and magnitude. The model postulates three main representations of numbers:

A verbal code in which numbers are represented as a parsed sequence of words.
A visual Arabic code in which numbers are represented as identified strings of digits.
An analogical quantity or magnitude code.

Each part of the model has been associated with increased activity in a particular part of the brain. For example, the horizontal segment of the intraparietal sulcus has been suggested as the region involved in encoding the analogical representation of numerical magnitude (Dehaene et al. 2003).

Transcoding Numerals (Symbols) to Number Words

Dehaene et al. (2003) proposed two major coordinated routes: a direct asemantic route that transcodes written numerals, i.e. the symbolic notations, to verbal representations and an indirect semantic route for quantitative processing. Alternative semantic routes are those that go through an intermediate step of activation of quantity associated with the target numeral.

In the working group 2 presentation and subsequent discussion, Gould raised the question of whether transcoding numerals to number words is limited to either semantic or asemantic pathways. Gould (2015) drew upon an example of a 7-year-old child's alternative strategy for locating and naming numerals on the number line. The discussion centred on a videoed interview with a child (Electronic Supplementary Material: Gould 2017), identified as Jed, indicating that these are not the only pathways used to transcode numerals to words. Instead of using a direct asemantic route to transcode written numerals to verbal representations, Jed used a more laborious transcoding pathway to identify numerals. Jed attempted to visualise the location of numerals on an ordered line of numbers. He would then seek the corresponding number word by counting from one. His process for identifying numerals was purposeful and time intensive, but clearly not asemantic. Gould's pertinent example shows that coding pathways associated with number may be more complex than the neuro-cognitive triple-code model currently allows. Whereas previously it was thought that Arabic numerals might activate representations of magnitude automatically, other research suggests that this is not the case (Rubinsten and Henik 2005). Gould proposes that learning to identify numerals is a learnt process. For most students, transcoding Arabic numerals to words is an asemantic process, but for some, it can rely upon a process that is not instant. For example, Frederick, a 7-year-old student in his second year of formal schooling, would regularly confuse 12 and 20. Over a period of 10 weeks, he learnt to correctly identify 12 and 20. However, to identify 12, that is, to say *twelve* in response to seeing it, Frederick counted from one to twelve subvocally.

Assessments of young children's counting often reveal the need to rely on the count from 1 (often subvocally) as the reliant strategy to 'reach' the required number name. These children are unable to move flexibly between different positions in the number sequence to count either forwards or backwards. When children move from producing an ordered sequence of counting words from 'one' to developing cardinal meaning, their understanding of the quantity is described as the count-to-cardinal transition (Fuson 1988, p. 266). What is important here is whether these children possess either cardinal or ordinal understanding of quantity or both, e.g. '12'. The findings highlighted by Sinclair and Coles (this chapter) raise further questions about ordinality – could Jed explain how numerals are ordered in a sequence? Another important observation is that the verbal processes concerned with learning the labels for Arabic numerals appear to be critical for arithmetic development between the ages of 6 and 7 years (e.g. see Göbel et al. 2014).

Beyond Neuro-cognitive Approaches: Quantitative Relations, SFOR and an Awareness of Patterns and Structures

This section provides an overview of current research that informs the research perspectives of working group 2. In this section, we will address the importance of reasoning about quantitative relations, children's spontaneous tendency to do so and their awareness of patterns and structures.

Children's Early Competencies in Quantitative Relations

It is apparent that the analysis of early mathematics-related competencies has capitalised on measures that emphasise children's numerical competencies, i.e. their subitising skills (Schleifer and Landerl 2011), counting skills (Geary et al. 1992), ability to compare numerical magnitudes (Griffin 2004) and ability to position numerical magnitudes on an empty number line (Siegler and Booth 2004). While such measures provided empirical evidence for the multicomponential nature and importance of young children's early numerical competencies (Dowker 2008), they imply a restricted view on children's early mathematical competencies and their importance for later mathematical development. Starting from Piaget's logical operations framework (e.g. Piaget and Szeminska 1952), there is a recent renewed research attention to children's quantitative reasoning skills, such as their understanding of the additive composition of number or their multiplicative reasoning skills, as well as to their importance for later mathematical learning at school (e.g. Clements and Sarama 2011; Nunes et al. 2008, 2012).

Several authors have explored the emergence and early development of these two forms of quantitative reasoning. As far as additive reasoning is concerned, various principles including the additive composition of number but also the commutativity ($a + b = b + a$), the addition-subtraction inverse ($a + b - b = a$) and the addition-subtraction complement principle ($a - b = . \rightarrow b + . = a$) have been intensively studied (Baroody et al. 2009; Bryant et al. 1999; Gilmore and Bryant 2006; Robinson et al. 2006), sometimes also in relation to children's actual use of these principles in their mental arithmetic (Baroody 1999; Peters et al. 2010). However, only few studies have explicitly addressed the question of how children's understanding of these principles affects their (later) achievement in whole number arithmetic. The limited available evidence from these few studies suggests that quantitative reasoning of this sort makes a specific contribution to achievement in whole number arithmetic. These studies highlight the importance of focusing on the relational aspects of quantitative reasoning as critical principles that contribute to strong mental computational flexibility.

Spontaneous Focusing on Numbers (SFON) and Quantitative Relations (SFOR)

The studies on the early development of children's quantitative reasoning reviewed in Section take a typically cognitive perspective. They hardly address children's attention to, and feeling for, quantitative relations. Quantitative reasoning in these studies often relates to multiplicative thinking. Recently, researchers have started to explore children's spontaneous focusing on quantitative relations (SFOR), as a follow-up of their investigations on children's spontaneous focusing on numerosity (SFON), which has already shown to have predictive power in explaining children's later mathematical achievement (Hannula and Lehtinen 2005). McMullen, Hannula-Sormunen and Lehtinen (McMullen et al. 2014) describe SFOR as 'the spontaneous (i.e. undirected) focusing of attention on quantitative relations and the use of these relations in reasoning' (p. 218).

A central idea underlying these SFON and SFOR tendencies is that there are individual differences not only in how learners reason about mathematics and use their numerical skills in learning or testing situations, wherein children are *guided* to the mathematical elements or relations in the situation, but also how often they *spontaneously* focus on mathematical aspects of informal everyday situations. In these situations, the recognition and use of quantitative aspects of the situation are done at the child's own initiative and thus undirected and spontaneous (e.g. Hannula and Lehtinen 2005; McMullen et al. 2013, 2014). Therefore, studies on SFON and SFOR do not examine whether learners *are able to* recognise or count exact number but rather whether they *spontaneously use* their available number recognition or quantitative reasoning skills in situations where they are not explicitly guided or instructed to do so.

An Integrated Perspective Focused on Patterns and Structures

Mulligan and Mitchelmore (2009) looked beyond research on early numeracy and single mathematical content domains such as counting to identify and explain common underlying bases of mathematical development. Drawing on their seminal studies of multiplicative reasoning and representations of number, they investigated the cognitive development of mathematics through the assessment of children's conceptual structures. A strong body of research on patterning, early algebraic thinking and the role of spatial structuring in mathematical representations supported their integrated theoretical approach that young children could develop cognitively sophisticated mathematical concepts. Based on a series of related studies with diverse samples of 4–8-year-olds, they identified and described a new construct, awareness of mathematical pattern and structure (AMPS), that generalises across mathematical concepts and processes and can be reliably measured (Mulligan and Mitchelmore 2013; Mulligan et al. 2015). Interestingly, just as McMullen et al. discuss, in their conceptualisation of AMPS, Mulligan and Mitchelmore (2009, p. 39) also look beyond children's ability in developing early numerical competence, by stating that AMPS may consist of 'two interdependent components: one cognitive (knowledge of structure) and one meta-cognitive, i.e., "spontaneous" (a tendency to seek and analyze patterns)'. Both are likely to be general features of how students perceive and react to their environment, according to these authors. In line with this construct of AMPS is the discussion in working group 5 on structure, also referring to the notion of structuring for mathematical competence and the work of John Mason.

Mathematical pattern involves any predictable regularity involving number, space or measure such as number sequences and geometrical patterns. Structure refers to the way in which the various elements are organised and related such as iterating a single 'unit of repeat' (Mulligan and Mitchelmore 2009). AMPS involves structural thinking based on recognising similarities and differences and relationships, but also a deep awareness of how relationships and structures are connected. Spatial structuring abilities provide the essential structural organisational features supporting numerical processes such as estimation of group size, multiplicative awareness of an array, iterating a unit of repeat in a repetition or equal partitions on a number line.

An interview-based assessment instrument was developed and validated, the *Pattern and Structure Assessment – Early Mathematics* (PASA) (Mulligan et al. 2015), across a wide range of concepts including patterning, spatial visualisation and early graphical representation. Responses included drawn representations and verbal explanations of patterns and relationships. Five broad levels of structural development were identified and described: prestructural, emergent, partial, structural and advanced structural (e.g. see Mulligan and Mitchelmore 2013). Further validation studies indicated that high levels of AMPS were correlated with high performance on standardised achievement tests in mathematics with young students (Mulligan et al. 2015). The PASA yields an overall AMPS score as well as scores on five individual structures (*sequences, shape and alignment, equal spacing, structured counting* and *partitioning*). All of these structures are highly interrelated. Repeating pattern sequences, equal spacing and structured counting all involve the idea of equal groups or units; shape and alignment considerations often result in equal groups; and partitioning requires the construction of equal groups or parts. In Chapter, a description of these structural groupings is provided in view of identifying common characteristics of AMPS that are often lacking in children with mathematics learning difficulties (MLD).

In alignment with the assessment of AMPS, an innovative, highly challenging alternative learning programme, the *Pattern and Structure Mathematics Awareness Program* (PASMAP), was developed and evaluated longitudinally in the kindergarten (the first year of formal schooling in Australia). This study provided the empirical evidence that young children are capable of representing, symbolising and generalising mathematical patterns and relationships, albeit at an emergent level (Mulligan et al. 2013). These findings suggest that restricting early learning to basic counting, simple arithmetic and informal notions of measure and geometry limits the development of AMPS. The study also tracked and described children's individual profiles of mathematical development, and these analyses showed that core, underlying mathematical concepts are based on AMPS and that some students develop these more readily and in more complex ways than others.

The PASMAP programme develops integrated learning experiences aimed at promoting visual memory, abstraction and generalisation, suitable for young students. Each PASMAP learning 'pathway' is directed mainly towards one or more of the five core structural groupings described above. The initial pathways include pattern as unit of repeat and growing patterns, grid structure, two-dimensional and three-dimensional relationships, structuring base ten, partitioning and sharing, equal grouping, unitising in measurement and symmetry and transformations. The first pathways are followed by more challenging tasks that link with the previous pathways and extend to multiplicative patterns, metric measurement, patterns in data and angles, direction and perspective taking. Clearly there is a strong thread of spatial structuring inherent in the pathways of learning.

In summary, there are strong connections among the theoretical approaches discussed in this section. They all highlight the importance of quantitative relations, patterns and structures as fundamental to whole number arithmetic. Recognising children's spontaneous attention to quantities can be linked to the development of AMPS that also focuses on children's natural tendency to seek structure in forming numerical relationships. In this respect, it is interesting to point to the paper presented by Sayers and Andrews (2015), which was presented in working group 1, but which also addressed the question of the foundations of number sense in a remarkably broad way. By summarising the recent research work in this domain, these authors arrived at a multidimensional framework, which they have called foundational number sense (FoNS), that comprises the following eight categories: number recognition, systematic counting, awareness of the relationship between number and quantity, quantity discrimination, an understanding of different representations of number, estimation, simple arithmetic competence and awareness of number patterns. This framework provides us, to some extent, with a comprehensive picture of early number competence that was also discussed in working group 2. The features that are not evident in this framework, i.e. awareness of mathematical patterns and structure (AMPS) and spontaneous focusing on number (SFON) and on relations (SFOR), are taken up in the working group 2 discussion.

Exemplars of Classroom Studies from Cognitive Perspectives

In this section, we describe some pertinent examples of intervention studies presented to working group 2 or other applications of the perspectives described in Section: ordinality (Sinclair and Coles 2015), the Italian *PerContare* project focused on structural relationships in arithmetic (Baccaglini-Frank 2015), improving early numeracy through additive relations (Roberts 2015), a cross-cultural study of kindergartners' number competence (Elia and van den Heuvel-Panhuizen 2015) and counting and representational structures (Milinković 2015).

Ordinal Awareness in Learning Number

For studies on ordinality, we draw on the work of Sinclair and Coles (2015). This raises an important question concerning typical developmental sequences posited by theories of early number learning, where what is given emphasis in the first years of schooling is training children to associate numbers with counting and matching sets of objects. Their research has led to the hypothesis that what is significant in the learning of number (and mathematics more generally) is not being able to link symbols to objects in a manner that is often considered accessible or natural but being able to link symbols to other symbols.

Sinclair and Coles (2015) make a distinction, in relation to number, between ordinal and cardinal aspects. They refer to *ordinality* as the capacity to place number words and numerals in sequence: for example, to know that 4 comes before 5 and after 3 in the sequence of natural numbers. Other aspects of ordinality, such as the use of ordinal names and symbols such as 'first' and '1st', may be common, for instance in the French tradition. Cardinality refers to the capacity to link number symbols to collections, e.g. to know that '4' is the correct representation to denote a group of four objects. They assert that the current emphasis on cardinal awareness in learning number may be misplaced (Coles 2014) and they have been exploring what is involved in developing greater ordinal awareness of number and what are the potential benefits?

Recent neuroscientific studies (e.g. Lyons and Beilock 2011) have challenged the dominant cardinal view of numerical cognition. Lyons and Beilock found a 'distance effect' persisted with the order comparison of groups of dots, but, importantly, when judging the order of numerals, the distance effect is reversed. In other words, when asked if three numerals are in order, the closer they are together, the quicker it is found that subjects can typically make the judgement of correct ordering or not. Lyons and Beilock used this reversal of the distance effect to suggest that the brain is doing something different when making ordinal comparisons of numerals, compared with both cardinal comparisons (of numerals or dots) and compared with ordinal comparisons of dots.

A common approach to working on ordinality in schools involves practising the number song; children are invited to count in ones up to 5 or 10, then 20 and then 100. While Sinclair and Coles see much value in this practice, as a first way of introducing children to the language and sounds of numbers, working on the successor function for integers does not exhaust the potential of ordinal awareness. This has already been made evident in the work of Gattegno (1974), whose curriculum for early number was based on developing awareness of relations among lengths, where what is symbolised are relations between objects (greater than, less than, double, half), rather than, say, using numerals to label 'how many' objects are in a collection. Gattegno introduced work on place value as a linguistic 'know-how' and not something that required 'understanding'. Similarly, the research discussed in working group 1 (Chap.) and also in Chap. refers to the important role of language in labelling numbers. Gattegno also made extensive use of fingers (both the teacher's and the children's) as haptic symbolic devices for working on number relations, with a focus on correspondence and complementarity. Sinclair and Coles see awareness of number, in this curriculum, arising out of linguistic skill and awareness of relations in a manner that does not emphasise nor require a cardinal focus on counting collections.

Sinclair and Coles direct attention to the importance of ordinality in researching the use, in the context of the early learning of number, of an innovative iPad app, *TouchCounts* (Sinclair and Jackiw 2011). The discussion in Chap. also refers to the use of *TouchCounts* as a tool for learning. *TouchCounts* was initially designed as a counting environment, to help children learn about one-to-one correspondence. Every time a finger touches the screen, a yellow disc appears, labelled with a numeral, and that numeral is spoken aloud. Each subsequent touch produces a yellow disc with the next numeral on it. With the gravity mode turned on, taps that are made by the child below the 'shelf' fall away, much in the same way that turning the page of a book makes that page number disappear. If one taps above the shelf, the yellow disc is 'caught' and remains on the shelf. It is thus possible to see just the yellow disc labelled '6' on the shelf if the previous five taps have been below the shelf. Notice that this task requires being aware of the fact that 5 comes before 6, but does not require any sense of cardinality. In both the temporal dimensions, but also because of the lack of cardinal reference, this Enumerating World emphasises ordinality. With the use of the aural feedback, as well as the numerals, there is also a strong emphasis on language and symbol, as per Lyons' recommendation (see Sinclair and Pimm 2015).

We draw upon this example of Sinclair and Coles from *TouchCounts* that points to the potential for ordinality in learning number (this volume, Sect.). In a kindergarten classroom, the children are sitting on the carpet, with the overhead projector hooked up to *TouchCounts*. The teacher has asked the children to count by 5s. They do this by tapping with four fingers (simultaneously) below the shelf and then once above. This leaves the multiples of 5 on the shelf. The children take turns doing the 4 + 1 tapping, but were asked to announce the number that would be on the shelf before starting tapping. Note that instead of hearing 'five, ten, fifteen…', the children hear 'four, five, nine, ten, fourteen, fifteen…'.

The teacher had intended to only get up to about 25, but the children wanted to keep going. At 125, they began to predict what number would appear on the shelf – chanting it out, chorus style – and ended up going all the way to 200. At this point, the following interaction took place:

Cam: I thought that two hundred was right after one hundred, but it's not.
Teacher: No, how far is it away from one hundred?
Cam: It's, it's, it's one more hundred away.

Significant in this episode is the fact that the children were involved in a skip-counting activity that had no explicit connection to a quantity of objects. Instead of seeing five objects as a cardinal quantity, they only saw the numbered object 5, as with all the multiples of 5. The attention was focused on the structure of the numbers, which is what enabled the children to begin to chant out the multiples. When the class reached 200, there had been no connection made between the number word and a quantity (of, say, two hundred objects). Indeed, Cam's realisation about the relation between 200 and 100 is not a cardinal one; he is instead basing his understanding of the relation on the observation that all the same multiples of 5 have to be done again in order to get from 100 to 200. In this sense, the relation seems to be deeply temporal, assembled as it is with the time it takes to create all the numbers up to 100 and then to 200. The relation is also entangled with *TouchCounts's* pronouncements ('one hundred', 'one hundred and forty-seven', 'two hundred'), some of which these children would never had heard before and which they could not have read from the symbolic forms (100, 147, 200), but could now begin to associate with those forms.

The discussion above suggests it is important to balance ordinal and cardinal aspects of number sense development in the primary grades. This will require some reflection on the ingrained ways in which cardinality is now privileged, as well as further creative explorations of how ordinality can be mobilised to promote the development of other number-related awareness such as place value.

Part-Whole Relations and Structure Sense

The Italian project (Baccaglini-Frank and Scorza 2013; Baccaglini-Frank and Bartolini Bussi 2015) builds upon a collaboration between cognitive psychologists and mathematics educators, aimed at developing teaching strategies for preventing and addressing early low achievement in arithmetic (also see papers by Young-Loveridge and Bicknell 2015, and Gervasoni and Parish (2015). This project takes an innovative approach to the development of number sense, that is, being grounded on a kinaesthetic and visual-spatial approach to part-whole relationships.

The project focuses on the importance of perceiving part-whole relationships and of becoming aware of structure (Baccaglini-Frank 2015; Electronic Supplementary Material: Baccaglini-Frank, 2017a). This demonstrates that part-whole relations arise from what Resnick and colleagues (Resnick et al. 1991) have described as proto-quantitative part-whole schemas 'that organize children's knowledge about the ways in which material around them comes apart and goes together' (p. 32). For example, part-whole thinking helps students recognise that numbers are abstract units that can be partitioned and then recombined in different ways to facilitate numerical calculation (Britt and Irwin 2011). Moreover, part-whole thinking is fundamental for higher mathematical reasoning. For example, the pre-algebra literature highlights how if attention is drawn to the development of part-whole relations, no longer do 'addition and subtraction appear as separate operations, but rather as dialectically interrelated actions that arise from the part-whole relation between quantities' (Schmittau 2011, p. 77).

Baccaglini-Frank (2015) refers to the part-whole relationship as a construct, highly resonant with the research on *awareness of mathematical pattern and structure* (AMPS) (Mulligan and Mitchelmore 2009, 2013). She refers to the critical feature of AMPS that is characterised by the child's structuring of groups to represent quantities; this involves part-whole relationships. Similarly, the ability to structure quantities is discussed in working group 5 (see Chap.) where the focus is on instruction that needs to provide opportunities for structural relationships to be associated with fundamental properties.

The following examples illustrate the use of hands and fingers to represent structure and the use of partitioning in the context of multiplication.

Hands and Fingers: An Important Embodied Structure

Various studies have highlighted how sensorimotor, perceptive and kinaesthetic-tactile experiences are fundamental for the formation of mathematical concepts – even highly abstract ones. For example, the key role attributed to the use of fingers in the development of number sense seems to be highly resonant with the frame of *embodied cognition*. Fingers and hands naturally embody part-whole relationships with respect to 5 and 10 and therefore can and should be used to foster such awareness. The didactical potential of hands and fingers in their natural positions can be exploited in many different ways (e.g. see Baccaglini-Frank 2015), including through multitouch technology, well before formal schooling starts (e.g. Baccaglini-Frank and Maracci 2015).

Use of Artefacts for Fostering the Development of Structure Sense: The Importance of Sharing Strategies

Various studies in mathematics education have focused on the design and implementation of didactical activities significantly based on bodily experience and on the manipulation of concrete objects with the aim of fostering the development of particular mathematical meanings. For example, in Chap. (Sect.), within a semiotic perspective, Bartolini Bussi and colleagues describe how the student's use of specific artefacts in solving mathematical problems contributes to his/her development of mathematical meanings, in a potentially 'coherent' way with respect to the mathematical meanings aimed at in the teaching activity (Bartolini Bussi and Mariotti 2008). However, they argue that it is important to keep in mind that in fostering the development of mathematical meanings an essential component is the students' sharing, comparing and evolving of strategies (which can be accomplished in a number of different ways). These mathematical meanings, of course, can include structure sense, and this can be promoted through a variety of different mathematical content.

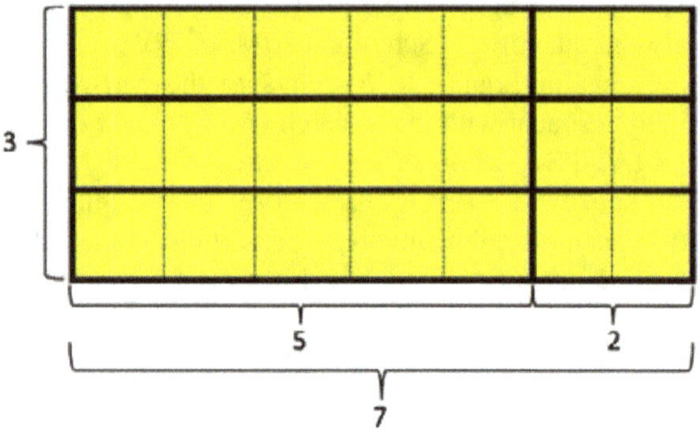

Fig. A possible decomposition of 7×3 into $5 \times 3 + 2 \times 3$

As an example, Baccaglini-Frank (2015) shows how 7-year-old students learned to think about (and perform) products (up to 10×10) within the Italian *PerContare* project. The children were introduced to *rectangle diagrams,* cardboard rectangles with a grid of 1cm^2 squares marking its dimensions, which represent the numbers to be multiplied. The area of the rectangle (and its unit of measure) is the number of squares that make it up. Various activities with the rectangles are proposed with the didactical goal of fostering the students' production of visual and kinaesthetic-tactile manipulative strategies for calculating products using number facts they already know. Typically, students in the experimental second grade classes already knew the sequences of the first 10 multiples of 1 (from counting), 2 (they had learned to 'double'), 5 (they could quickly add 'hands') and 10 (they could quickly count up bundles of straws). So the activities aimed at developing strategies of decomposition and composition based on such knowledge. For example, to calculate 7×3, children could think of 7 as $5 + 2$ and use the known rectangles 5×3 and 2×3 to build the total rectangle. Figure shows an example.

The different strategies used by the children were compared and discussed. By the end of the school year, many children were able to perform calculations without the support of the physical rectangle diagrams any longer. For example, below is the verbal description produced by Marco (7 years 8 months) of the mental (and highly visual) strategy he uses to figure out 7×8 when he is called on by the teacher.

Teacher: Without drawing the 'building[2] seven times eight, can you tell me how you break it and count it?

Marco: So, seven times eight… I break it into five and two, and I count it: five, ten, fifteen, twenty, twenty-five, thirty, thirty-five, forty… and I already have forty. Then I count the twos: two, four, six, eight, ten, twelve, fourteen, sixteen. Then I do forty plus, uh, I break the sixteen into ten and six, and I do forty plus ten, fifty, then I add those six and it is fifty-six.

Teacher: Wow! You are tremendous!

[2] In this class the teacher took on the class' idea to refer to the rectangle diagrams as (apartment) 'buildings' which could be split and put back together.

As he speaks, Marco frequently gazes into space, as if he were seeing the diagram he is decomposing and recomposing. Mulligan et al. (2013) refer to this as visualising the structure, a central component of AMPS.

The examples provided by Baccaglini-Frank (2015) turn our attention to the critical role of the structure of artefacts and the ways that young students interpret and construct representations. In the case of Marco, he has internalised the visualised 'structure' of the diagram, and we can infer that he had internalised the structure of the grid. The use of structure sense is embedded within this example. The ability to decompose or partition mathematical representations is directly linked to the child's strategies for calculating, often articulated by the child's strong visual imagery of buildings to be broken up and through verbalisation of 'I break… into parts'. The key process here is not counting by ones or repeated addition but structuring by partitioning or 'breaking up'. Some knowledge of base-ten structure is also evident here. Here we see similarities with the work of Young-Loveridge and Bicknell in their paper discussing the role of structure in terms of place value and grouping (working group 3, Chapter of this volume).

Additive Relations

The study by Roberts (2015) complements the work of Baccaglini-Frank (2015) pertaining to part-whole and structure sense. Roberts presented to working group 2 a conceptual framework for interpreting children's external representations of whole number additive relations in the early grades (Roberts 2015). She bases her approach on growing evidence from classroom studies in South Africa that one of the major factors inhibiting learners' mathematical progression is continued using of counting by ones strategies for mathematical calculations. This concern is not exclusive to studies with South African teachers and young children; for example, Young-Loveridge highlights the same concern in her studies from New Zealand reported in Chap.(this volume).

Roberts explores young learners' representations of additive relationships and provides insights into underlying structure linked to grouping. She presents an adapted framework focusing on shifts within modes of representation that denote a move from counting to calculating. Based on the work of Ensor et al. (2009), she presents the adapted framework moving from concrete apparatus to iconic images, to indexical images (generic), to symbolic (number-based) and to the abstract symbolic-syntactical level. What the study suggests is that progression within these dimensions varies when various modes of representation are used for different tasks and at different times over the 10-day intervention period. What Roberts proposes is that it is important for teachers to attend to both structure (arrangement and group-wise) and action within a particular mode of representation, when interpreting learners' representations of additive relations.

Roberts' approach that articulates the need for children to work flexibly with multiple modes of representation is exemplified; particular representational types are not automatically mapped to a particular calculation strategy. How the interplay of the various modes of representation interrelates with the developing sense of structure and the complexity of the structural features of the tasks at hand is exemplified. However, what we see here is the attempt to integrate complex aspects of structural development with more traditional, broad levels of progression from concrete to abstract thinking, as well as consideration of embodied action. What is difficult to assess is whether the structure dimensions direct or dominate the influence of other dimensions. There is clearly emphasis on the use of colinearity (left-right, top-bottom), linear directions and partitioning. This complex matrix approach raises relevant research questions about how the internalisation of the structural features occurs within and across the four dimensions over time and how this promotes abstraction and generalisation in developing arithmetic relationships such as equivalence or commutativity.

Cross-Cultural Study of Number Competence

Elia and van den Heuvel-Panhuizen investigated the number competence of kindergartners from the Netherlands ($n = 334$) and Cyprus ($n = 304$). The study supported the multidimensional nature of kindergartners' number development. Although the study did not include assessment items from the full domain of number and operations, four structures were found to be central to number competence: counting, subitising, additive and multiplicative reasoning. The children from the Netherlands outperformed those from Cyprus, demonstrating competence across the four components. The number competence of the children from Cyprus reflected two components, including extended counting and additive reasoning. The discussion focused on possible reasons for differences in competence where it was considered that the Cyprus kindergarten's mathematics curriculum and teaching practices may have been restricted to counting and additive reasoning and less attention was placed on subitising and multiplicative reasoning. Counting strategies may have dominated the Cyprus children's strategies. It was clear that young kindergarten children could solve multiplicative items and they connected multiplication and divisions processes, although the multiplicative items were the most difficult. This was consistent with other studies presented to working group 2.

Mulligan and colleagues reported similar findings to those of the Netherlands sample in their assessment of counting, subitising and multiplicative reasoning with kindergarten students from Australia using the interview-based Pattern and Structure Assessment (PASA). The possible over-reliance on counting was also described in the studies by Roberts and Milinković. Here we refer to the study of Gervasoni and Parish (2015), where they used individual interviews to assess over 2000 Australian primary-aged children from grades 1–4. Counting, place value, additive and multiplicative tasks were administered with gradual increase in competence, but an over-reliance on counting strategies even at grade 4 level was found. The working group 2 also questioned the limitations of some early number competency assessments that may restrict items to counting and additive processes and argued for more assessments to probe mental calculation strategies.

Counting and Representations of Number

Milinković examined the development of young Serbian children's initial understanding of representations of whole numbers and counting strategies in a large cross-sectional study of 661 children aged 3–7 years. Individual interviews were conducted with a consistent set of 24 tasks across the sample so that developmental patterns in performance based on age categories could be ascertained. Although many of the tasks replicated a traditional approach, such as focusing on counting, set representation and one-to-one correspondence, there were some tasks that focused on structure through different spatial arrangements of groups of objects. Further there were tasks that required children to complete a two-dimensional drawing to show a quantity (box task) and to extend numbers on a number line that did not include equidistant points to assist in this process.

Although the research report is limited to performance data across the sample, there are some critical features inherent in some tasks that relate to other research discussed in working group 2. Milinković highlights the analysis of children's understanding of different graphical representations – the box diagram and number line – and presents some pertinent examples. Representations such as sets and the number line were found to be limited in their recordings. The ability to use equal spacing or a composite unit (equal size) to represent number appeared most difficult in the sequence of tasks developmentally.

Methodological Issues and Recommendations

In this section, we examine methodologies utilised in the type of studies on children's whole number learning and development being reviewed in this section. Evidently, these methodologies do not cover the whole range of research methods being used in the domain of WNA. Rather than providing a broad overview of the topic, we focus on two issues. First, we discuss study designs and their potentialities and limitations for understanding how children develop competencies with whole numbers. The discussion is restricted to cross-sectional, longitudinal and intervention studies. Second, we discuss task designs in cognitive neuroscience research pertinent to number learning. Here, we focus on the validity of tasks that are common to many studies measuring the understanding of magnitude of numbers. Such discussion reflects some of the issues raised in the previous ICMI Study 22 in 2014 (Watson and Ohtani 2015). To illustrate methodological issues, we refer to key aspects of whole number learning, such as strategy use, developmental aspects and the effectiveness of instructional approaches. We conclude with some recommendations for further research on whole number learning.

Study Designs

Assessing Strategy Use with Cross-Sectional Studies

Whole number learning requires, among other things, learning increasingly advanced strategies. For example, young children might initially use counting strat-egies to solve addition tasks, but they might later use stepwise addition strategies or they might be able to retrieve the results from memory by retrieval of known num-ber facts. Cross-sectional studies allow insights into children's performance and strategy use at a particular point in time. They also allow for investigation of how performance and strategy use depend on specific types of tasks and how performance and strategy use vary between students with different educational and sociocultural backgrounds (e.g. He 2015; Ma et al. 2015; Milinković 2015; Verschaffel et al. 2015; Yang 2015). Chapter on language aspects, Chap. on reporting the discussion of working group 1 and the commentary paper by David Pimm (Chap.) each address the role of language and culture, also from a historical perspective in the development of whole number arithmetic. Here, we turn attention to the work-ing group 1 discussion that shows how differences in number names according to culture may lead to wide differences in learning and pedagogical strategies.

An important issue in studying strategy use is the interplay between individual strategy use, individual ability and the affordances of a specific task. For example, children might not always use the most sophisticated strategy they could possibly use if less sophisticated strategies are more efficient for the specific task at hand. On the other hand, students might not be able to adapt their known strategy to the specific task. For that reason, researchers have argued that if in the assessment situation students are allowed to select their preferred strategy, we cannot draw valid conclusions concerning strategy efficiency (Siegler and Lemaire 1997). To draw such conclusions, it is necessary to compare students' performance in a choice condition wherein they are free to select their preferred strategy to their performance in a no-choice condition wherein students are forced to use a particular strategy. Many researchers have used this choice/no-choice method to study strategy efficiency and strategy flexibility (e.g. Verschaffel et al. 2015). This line of research has produced interesting and sometimes surprising results. For example, it seems that students do not always use the most efficient strategies that they may have acquired through instruction at school. Students also rely on strategies they have not been taught, and they might even invent their own strategies.

There is a rich and diverse range of studies that have examined strategy development and strategy use in early arithmetic development (e.g. see the papers by He on Chinese students' cognitive strategies to addition/subtraction problems (He 2015) and Yang on students' ability to judge the reasonableness of computational strategies (Yang 2015)). Many studies on strategy use have also focused particularly on promoting numeracy programmes or frameworks. However, to date, the complex interplay between the factors that influence strategy use on WNA tasks is not fully understood.

An important limitation of cross-sectional studies is that they do not allow conclusions to be drawn with respect to individual development or causal relations to be determined between foundational or natural abilities and mathematical learning. For that purpose, we need longitudinal studies, intervention studies and those that track, possibly from the origins, the growth of individuals' strategy development in WNA.

Tracing Individual Development with Longitudinal Studies

Longitudinal studies rely on data from individual children assessed over a longer period of time. In the case of numerical development, such studies allow for identifying those variables assessed early in development that are most predictive of later arithmetic achievement. While longitudinal studies on arithmetic development have been relatively scarce until two decades ago, an increasing number of longitudinal studies have been carried out since (e.g. see the above-mentioned synthesis of research on early number sense by Sayers and Andrews 2015). Many of these studies have produced converging results. For example, several studies found that at pre-school age, counting and linking quantity to number words are important predictors of mathematical achievement in the first years of primary school (Aunio and Niemivirta 2010; Krajewski and Schneider 2009). Other researchers have combined several measures basic understanding of numbers with the concept of number sense. Number sense, measured at the beginning of schooling, predicted achievement in school mathematics in the first and third grades (Jordan et al. 2010). In a 6-year longitudinal study, Reeve, Reynolds, Humberstone and Butterworth (2012) clustered children at the age of 6 years according to their basic numerical abilities such as dot enumeration and number comparison. The authors found that the clusters were relatively stable over the period of the study and that membership to a cluster was a robust predictor of arithmetic ability 5 years later.

Although longitudinal studies have contributed to our understanding of how certain arithmetic abilities develop over time, most of these studies have focused strongly on cognitive variables related to mathematics while paying less attention to general cognitive variables (such as IQ and working memory) or environmental variables (such as school environment, classroom teaching or socio-economic variables) (but see Skwarchuk et al. 2014). From a mathematics education point of view, this is problematic, because these more distal variables might strongly influence children's development.

The benefit for mathematics education of identifying the most relevant early predictors of arithmetic competencies is that we can develop teaching approaches that specifically address these predictors. Yet, we need further research to evaluate whether the developed teaching approaches are actually effective and to identify the most effective one(s) among competing teaching approaches. Intervention studies are suitable for that purpose.

Evaluating Teaching Approaches with Intervention Studies

Intervention studies have the advantage that, if properly designed, they allow conclusions about the causal effects of specific factors. This is at least the case when the intervention conditions are highly similar with respect to non-relevant factors. When designing an intervention study, an important question is how to design the control group condition. The challenge is that the question of what we want the experimental condition to be compared with to is not always obvious. Consider a game-based intervention study in which the experimental group uses a computer game that includes carefully designed number tasks. As a control condition, one might want to vary the method of instruction (computer-based versus not computer-based), the specific tasks (innovative tasks versus traditional tasks), the entertaining nature of instruction (game versus no game), the instructional setting (collaboration versus individual) or other factors. However, it is often impossible to vary all these factors within the same study. In addition, there might be theoretical reasons why combining certain factors is not reasonable from a mathematics education point of view. For example, collaborative learning might be more reasonable when the students work on problem-solving tasks than when they try to memorise arithmetic facts. Moreover, instructional factors are often closely related to one another, so that manipulating one factor can affect another factor.

Although strictly controlling the intervention conditions is necessary to draw conclusions about the causal effects of specific factors, doing so might reduce the external or ecological validity of the study. The reason is that the effectiveness of teaching approaches under controlled conditions might not transfer to regular, much more complex learning situations. Ideally, we need both highly controlled intervention studies and less strictly controlled classroom evaluation studies in order to compensate for the disadvantages of each. This will require replicating studies in a variety of settings and combining a variety of research methods (Schoenfeld 2007; Stokes 1997).

Task Designs

As discussed earlier in the first section of this chapter, neuroscience studies have addressed the brain mechanisms that underlie number processing. A main conclusion from this research field is that the human brain seems well prepared for processing (numerical) magnitudes. Although understanding number magnitudes has been a matter of research long before neuroscience studies identified the relevant brain areas, this conclusion increased the attention researchers paid to processing numerical magnitudes. Likewise, although the relation between mathematical abilities and other cognitive abilities has been studied for a long time, the fact that the intraparietal sulcus is a brain region responsible for magnitude processing as well as spatial thinking has influenced a number of studies addressing the relationship between numerical and spatial abilities (e.g. Mulligan and Woolcott 2015).

Although there is no doubt that understanding numerical magnitudes is an important facet of whole number arithmetic abilities, its particular role for arithmetic development is not completely clarified yet. One reason among others is that the tasks (measures) that have been used in previous studies for assessing magnitude understanding may not be as valid as many have thought them to be. Most studies have used number comparison tasks or number line estimation tasks to assess magnitude understanding. In number comparison tasks, one has to decide which of two numbers is numerically larger. In number line estimation tasks, one has to place a given number in the correct position on an empty number line. Performance on both the number comparison task and the number line estimation task has proven to be highly predictive of mathematical learning (e.g. Booth and Siegler 2008). Many researchers have concluded that processing numerical magnitudes is essential for learning of numbers and they have used either task to assess magnitude understanding. Surprisingly, studies that addressed the relation between different measures challenged the assumption that these different measures rely on the same cognitive mechanisms. Studies documented that the correlation between performance on number comparison tasks and number line estimation tasks was very small (Sasanguie and Reynvoet 2013) and that performance on symbolic and non-symbolic number comparison tasks was virtually unrelated (Gilmore et al. 2011). Meanwhile, there is converging evidence that the association between number comparison and arithmetic competence is much stronger for symbolic than non-symbolic measures (Sasanguie et al. 2014; Schneider et al. 2017), suggesting that it is the proficient use of number symbols that has a strong association with arithmetic competence.

Recent studies question the assumption that number comparison and number line estimation are 'pure' measures of magnitude understanding. A possible explanation could be that depending on the specific numbers involved, these tasks can be solved by strategies that vary in how strongly they require magnitude understanding. For example, for comparing two-digit numbers, one can rely on digit-by-digit comparison without taking into account the magnitudes of the numbers as a whole. Likewise, researchers have argued that non-symbolic number comparison tasks may not only measure magnitude understanding as such, but also the capacity to suppress irrelevant visual cues (Clayton and Gilmore 2015) or to switch the focus on reliable cues (Gebuis and Reynvoet 2012). For number line estimation tasks, the strategies one can use also depend strongly on the specific numbers. Finding the correct position of 50 on a number line from 0 to 100 is easy (because finding the midpoint of the line is a visually simple task), while finding the correct position of 83 is more difficult, because there is no clear benchmark that can be used. Recent research has documented that already second grade children use a variety of strategies for solving number line estimation tasks and that these strategies depend on modes of presentations and the availability of benchmarks (Peeters et al. 2015).

More generally, most cognitive and neuroscience studies have used very simple tasks. For example, studies on the neural underpinnings of mental arithmetic have often used single-digit addition tasks. Although these are important first steps, and although performance of these tasks might be related to mathematical achievement later on, studying these tasks is not sufficient to explain mathematical thinking, which is typically much more complex. Accordingly, we should be cautious about interpreting neuroscience findings in terms of educational implications. In particular, we should not consider neuroscientific data as more convincing or informative than behavioural data (Beck 2010; De Smedt et al. 2011).

Conclusions: Methodological Issues

Some tentative conclusions and implications can be gleaned from the discussion provided in the above section. Studies that aim to provide predictive factors for math-related competencies may need to take into account the influence of a broader range of variables such as IQ, working memory and socio-contextual factors. The analyses of specific tasks such as number line estimation do not reflect the wide variation in children's own strategies that they may impose on the tasks. For example, the type and size of numbers may vary, but solutions may also depend on modes of presentation and the availability of benchmarks. Most cognitive and neuroscience studies have used very simple tasks limited to one area of competence. The limitations of these findings should be acknowledged in view of the much more complex relationships between concepts and processes that contribute to mathematics learning and thinking. Further, we need longitudinal studies to understand better how numerical abilities develop over time.

General Conclusions and Implications

General Conclusions

This chapter has highlighted the need to review neuro-cognitive, cognitive and developmental approaches to number learning and the measurement of numerical abilities. While critical components of WNA may differ between disciplines, some important commonalities have been found between approaches. Butterworth's research (2015) focused on the 'starter kit' for number also reflected conceptual foci on studies on early number from a cognitive mathematics education perspective. Tasks that incorporate subitising and numerical estimation are common to mathematics education psychological studies but differ methodologically. Although these neuro-cognitive studies provided convincing evidence of specific features of early number development, these were limited to numerical magnitude and cardinality, assessed in clinical studies. Cognitive studies on early numerical and general mathematical competencies have received inadequate attention in the neuroscience field.

From various cognitive perspectives, key components and processes integral to mathematics learning and related to WNA were described: spatial reasoning and spatial sense, reasoning about quantities and relationships, SFON and SFOR, structural relations and patterns (AMPS), ordinality, partitioning and representing numerical relationships. These studies, together, provide evidence that young children are capable of quantitative reasoning from a young age. In particular, Section highlighted recent research on young children's capacity to reason about quantitative relations (SFOR), as well as their *spontaneous tendency* to do so. This line of research shows strong synergies with the structural approach to early number development that focuses on awareness of mathematical patterns and structures (AMPS). In their conceptualisation of AMPS, Mulligan and Mitchelmore also go beyond the idea of early numerical competence based on ability. AMPS consists of two interdependent components: one cognitive (knowledge of structure) and one metacognitive (a tendency to seek and analyse patterns). It seems that reasoning about quantities and relationships, SFON and SFOR, structural relations and patterns (AMPS), ordinality, partitioning and representing numerical relationships are related to structural development in mathematics.

The exemplars of studies presented in Section each reflect the need to take a more integrated approach to early WNA development. The studies point to a common approach that seeks to reveal the deep interconnected structural features of conceptual development of number. The studies of Baccaglini-Frank and colleagues turn attention to the critical role of the structure of artefacts and the ways that young students interpret and construct representations. The use of structure sense is embedded within most examples. The ability to decompose or partition mathematical representations is also featured in the work of Milinković and colleagues. Several studies reflected that importance of varying models of representation supporting the idea of complexity of the learning process that is often absent from neat theoretical frameworks of likely learning progression. The discussion of methodological issues in Sect. raises questions for future research and practice.

There were several papers in this working group which focused on learners with special needs: Butterworth (2015) drew attention to the prevalence and diagnosis of dyscalculia; Baccaglini-Frank's paper (2015) reported on an intervention study in Italy designed to redirect Italian learners at risk of dyscalculia diagnosis; Gould's study (2015) focused on one child who used an atypical way of counting; and the Roberts (2015) paper was motivated by the prevalence of South African children aged 10–12 years using such inefficient unit counting strategies (long after this was developmentally appropriate). These papers depicted groups and individual children not progressing mathematically as expected in relation to their peers and/or the national mathematics curriculum. Issues concerning children with special learning needs are discussed further in Chapter.

Implications for Further Research and Practice

The chapter has highlighted three issues that lead to recommendations for further research. Firstly, we need longitudinal studies to better understand how numerical abilities develop over time. These studies should take into account not only math-related variables but also other variables that have crucial effects on development, such as IQ, working memory and contextual variables such as socio-economic factors and school environments. This would be helpful in putting the relevance of specific, math-related predictors into perspective. Although cross-sectional studies can hint at relevant relationships between specific sub-competencies, only longitudinal studies support conclusions about children's individual development and the causal relations between foundational math-related abilities and whole number arithmetic learnt at school.

Second, there is a need for intervention studies to develop evidence-based instructional tasks, tools and techniques. This would enhance educational practice and also contribute to our better understanding of the causal effects of arithmetic development. Combining both strictly controlled experimental studies and less strictly controlled field studies allows us to overcome the drawbacks of each (Schoenfeld 2007).

Third, we need more studies that systematically evaluate the validity of simple number tasks that have been used frequently in (neuro-)cognitive studies. A multi-method approach (as used in the study by Peeters et al. 2015) seems promising for that purpose. Once appropriate tasks and methods are available, future neuroscience studies could address more complex mathematical thinking.

A less (neuro-)psychologically dominated and more interdisciplinary approach might bring a broader, more balanced perspective that takes into account both empirically based and classroom-oriented research from cognitive and developmental views of WNA. Classroom intervention studies do not easily permit generalisation, nor do they reflect the highly controlled experimental settings of the neuro-cognitive studies, but these studies are critical to informing further research, mathematics teaching practice and curriculum development.

Several chapters in the volume have centred on teaching practices and tools for learning whole number arithmetic. For example, working group 3 (Chap.) discusses cultural artefacts and tasks and working group 4 (Chap.) teaching and assessment approaches. While this chapter has delved into the cognitive and neuro-cognitive bases of research related to concept development in number, there are clearly synergies between this chapter and teaching approaches.

Some important messages emanating from the working groups have been articulated for teachers so that they develop their professional knowledge and improved awareness of the complexities of whole number learning. Cognitive and neuro-cognitive approaches can enable new insights to be incorporated into teaching practices. Aligned with new insights is the need for effective professional learning programmes to enable teachers to implement and review new approaches, tasks or assessment practices that they adopt. Developing a better understanding of the wide variations in students' strategies and the difficulties students experience in acquisition of number concepts is critical to improving mathematics learning overall.

The research discussed in this chapter can provide to some extent explanations and possible interventions to assist teachers to focus on core mathematical foundations such as numerical magnitude representation and the mental number line, structures and relationships in developing number sense, promoting multiplicative thinking rather than restricting focus on counting and additive computations and attention to the role of spatial origins of number learning. The role of students' representations and interpretations of those representations has been exemplified. These examples may assist teachers in selecting appropriate representational tools and tasks to promote better understanding of whole number relationships. In conclusion this chapter has raised new questions from a range of perspectives, both neuro-cognitive and cognitive, but with a common goal of providing new insights into the complex and dynamic nature of young students' whole number learning.

References

Aunio, P., & Niemivirta, M. (2010). Predicting children's mathematical performance in grade one by early numeracy. *Learning and Individual Differences, 20*, 427–435.

Baccaglini-Frank, A., & Bartolini Bussi, M. G. (2015). Buone pratiche didattiche per prevenire falsi positivi nelle diagnosi di discalculia: Il progetto PerContare. *Form@re-Open Journal per la formazione in rete, 15*(3), 170–184.

Baccaglini-Frank, A., & Maracci, M. (2015). Multi-touch technology and preschoolers' development of number-sense. *Digital Experiences in Mathematics Education, 1*(1), 7–27.

Baccaglini-Frank, A., & Scorza, M. (2013). Preventing learning difficulties in early arithmetic: The PerContare Project. In T. Ramiro-Sànchez & M. P. Bermùdez (Eds.), *Libro de Actas I Congreso Internacional de Ciencias de la Educatiòn y des Desarrollo* (p. 341). Granada: Universidad de Granada.

Baroody, A. J. (1999). Children's relational knowledge of addition and subtraction. *Cognition and Instruction, 17*(2), 137–175.

Baroody, A. J., Torbeyns, J., & Verschaffel, L. (2009). Young children's understanding and application of subtraction-related principles. *Mathematical Thinking and Learning, 11*(1–2), 2–9.

Bartolini Bussi, M. G., & Mariotti, M. A. (2008). Semiotic mediation in the mathematics classroom artefacts and signs after a Vygotskian perspective. In L. English (Ed.), *Handbook of international research in mathematics education* (2nd ed., pp. 746–783). New York: Routledge.

Beck, D. M. (2010). The appeal of the brain in the popular press. *Perspectives on Psychological Science, 5*, 762–766.

Booth, J. L., & Siegler, R. S. (2008). Numerical magnitude representations influence arithmetic learning. *Child Development, 79*, 1016–1031.

Britt, M. S., & Irwin, K. C. (2011). Algebraic thinking with and without algebraic representation: A pathway for learning. In J. Cai & E. Knuth (Eds.), *Early algebraization: A global dialogue from multiple perspectives* (pp. 137–160). Berlin: Springer.

Bryant, P., Christie, C., & Rendu, A. (1999). Children's understanding of the relation between addition and subtraction: Inversion, identity and decomposition. *Journal of Experimental Child Psychology, 74*, 194–212.

Butterworth, B. (2005). The development of arithmetical abilities. *Journal of Child Psychology and Psychiatry, 46*(1), 3–18.

Butterworth, B., & Reeve, R. (2008). Verbal counting and spatial strategies in numerical tasks: Evidence from indigenous Australia. *Philosophical Psychology, 21*, 443–457.

Clayton, S., & Gilmore, C. (2015). Inhibition in dot comparison tasks. *ZDM Mathematics Education, 47*(5), 759–770.

Clements, D. H., & Sarama, J. (2011). *Tools for early assessment in math (TEAM). Teacher's Guide*. Columbus: McGraw-Hill Education Series.

Coles, A. (2014). Ordinality, neuro-science and the early learning of number. In P. Liljedahl, C. Nicol, S. Oesterle, & D. Allan (Eds.), *Proceedings of the Thirty-eighth annual conference of the International Group for the Psychology of Mathematics Education* (Vol. 2, pp. 329–336). Vancouver: PME.

De Smedt, B., Ansari, D., Grabner, R. H., Hannula-Sormunen, M., Schneider, M., & Verschaffel, L. (2011). Cognitive neuroscience meets mathematics education: It takes two to tango. *Educational Research Review, 6*, 232–237.

Dehaene, S. (2011). *The number sense: How the mind creates mathematics*. New York: Oxford University Press.

Dehaene, S., Piazza, M., Pinel, P., & Cohen, L. (2003). Three parietal circuits for number processing. *Cognitive Neuropsychology, 20*, 487–506.

Dowker, A. (2008). Individual differences in numerical abilities in preschoolers. *Developmental Science, 11*, 650–654.

Ensor, P., Hoadley, U., Jacklin, H., Kuhn, C., Schnitte, E., Lombard, A., & van den Heuvel-Panhuizen, M. (2009). Specialising pedagogical text and time in Foundation Phase numeracy classrooms. *Journal of Education, 47*, 5–30.

Fuson, K. C. (1988). *Children's counting and concepts of number*. New York: Springer-Verlag.

Gattegno, C. (1974). *The common sense of teaching mathematics*. New York: Educational Solutions Worldwide Inc..

Geary, D. C., Bow-Thomas, C. C., & Yao, Y. (1992). Counting knowledge and skill in cognitive addition: A comparison of normal and mathematically disabled children. *Journal of Experimental Child Psychology, 54*, 372–391.

Gebuis, T., & Reynvoet, B. (2012). The interplay between nonsymbolic number and its continuous visual properties. *Journal of Experimental Psychology: General, 14*, 642–648.

Gilmore, C., & Bryant, P. (2006). Individual differences in children's understanding of inversion and arithmetical skill. *British Journal of Educational Psychology, 76*, 309–331.

Gilmore, C., Attridge, N., & Inglis, M. (2011). Measuring the approximate number system. *Quarterly Journal of Experimental Psychology, 64*, 2099–2109.

Göbel, S. M., Watson, S. E., Lervåg, A., & Hulme, C. (2014). Children's arithmetic development: It is number knowledge, not the approximate number sense, that counts. *Psychological Science, 25*(3), 789–798.

Griffin, S. (2004). Teaching number sense. *Educational Leadership, 61*(5), 39–42.

Hannula, M. M., & Lehtinen, E. (2005). Spontaneous focusing on numerosity and mathematical skills of young children. *Learning and Instruction, 15*(3), 237–256.

Jordan, N. C., Glutting, J., & Ramineni, C. (2010). The importance of number sense to mathematics achievement in first and third grades. *Learning and Individual Differences, 20*, 82–88.

Krajewski, K., & Schneider, W. (2009). Early development of quantity to number-word linkage as a precursor of mathematical school achievement and mathematical difficulties: Findings from a four-year longitudinal study. *Learning and Instruction, 19*(6), 513–526.

Lyons, I., & Beilock, S. (2011). Numerical ordering ability mediates the relation between number-sense and arithmetic competence. *Cognition, 121*(2), 256–261.

McMullen, J., Hannula-Sormunen, M. M., & Lehtinen, E. (2013). Young children's recognition of quantitative relations in mathematically unspecified settings. *Journal of Mathematical Behavior, 32*, 450–460.

McMullen, J., Hannula-Sormunen, M. M., & Lehtinen, E. (2014). Spontaneous focusing on quantitative relations in the development of children's fraction knowledge. *Cognition and Instruction, 32*(2), 198–218.

Mulligan, J. T., & Mitchelmore, M. C. (2009). Awareness of pattern and structure in early mathematical development. *Mathematics Education Research Journal, 21*(2), 33–49.

Mulligan, J. T., & Mitchelmore, M. C. (2013). Early awareness of mathematical pattern and structure. In L. English & J. Mulligan (Eds.), *Reconceptualizing early mathematics learning* (pp. 29–46). Dordrecht: Springer.

Mulligan, J. T., Mitchelmore, M. C., English, L., & Crevensten, N. (2013). Reconceptualizing early mathematics learning: The fundamental role of pattern and structure. In L. English & J. Mulligan (Eds.), *Reconceptualizing early mathematics learning* (pp. 47–66). Dordrecht: Springer.

Mulligan, J. T., Mitchelmore, M. C., & Stephanou, A. (2015). *Pattern and Structure Assessment (PASA): An assessment program for early mathematics (Years F–2) teacher guide*. Melbourne: ACER Press.

Nunes, T., Bryant, P., Burman, D., Bell, D., Evans, D., & Hallett, D. (2008). Deaf children's informal knowledge of multiplicative reasoning. *Journal of Deaf Studies and Deaf Education, 14*, 260–277.

Nunes, T., Bryant, P., Barros, R., & Sylva, K. (2012). The relative importance of two different mathematical abilities to mathematical achievement. *British Journal of Educational Psychology, 82*, 136–156.

Peeters, D., Degrande, T., Ebersbach, M., Verschaffel, L., & Luwel, K. (2015). Children's use of number line estimation strategies. *European Journal of Psychology of Education*. (Online first) doi:https://doi.org/10.1007/s10212-015-0251-z

Peters, G., De Smedt, B., Torbeyns, J., Ghesquière, P., & Verschaffel, L. (2010). Using addition to solve subtractions in the number domain up to 20. *Acta Psychologica, 133*, 163–169.

Piaget, J., & Szeminska, A. (1952). *The child's conception of number*. London: Routledge. (Original work published 1941).

Reeve, R., Reynolds, F., Humberstone, J., & Butterworth, B. (2012). Stability and change in markers of core numerical competencies. *Journal of Experimental Psychology: General, 141*, 649–666.

Resnick, L. B., Bill, V. L., Lesgold, S. B., & Leer, N. M. (1991). Thinking in arithmetic class. In B. Means, C. Chelemer, & M. S. Knapp (Eds.), *Teaching advanced skills to at-risk students* (pp. 27–53). San Francisco: Jossey-Bass.

Robinson, K. M., Ninowski, L. E., & Gray, M. L. (2006). Children's understanding of the arithmetic concepts of inversion and associativity. *Journal of Experimental Child Psychology, 94*, 349–362.

Rubinsten, O., & Henik, A. (2005). Automatic activation of internal magnitudes: A study of developmental dyscalculia. *Neuropsychology, 19*, 641–648.

Sasanguie, D., & Reynvoet, B. (2013). Number comparison and number line estimation rely on different mechanisms. *Psychologica Belgica, 53*, 17–35.

Sasanguie, D., Defever, E., Maertens, B., & Reynvoet, B. (2014). The approximate number system is not predictive for symbolic number processing in kindergarteners. *The Quarterly Journal of Experimental Psychology, 67*, 271–280.

Schleifer, P., & Landerl, K. (2011). Subitizing and counting in typical and atypical development. *Developmental Science, 14*, 280–291.

Schmittau, J. (2011). The role of theoretical analysis in developing algebraic thinking: A Vygotskian perspective. In J. Cai & E. Knuth (Eds.), *Early algebraization: A global dialogue from multiple perspectives* (pp. 71–86). Berlin: Springer.

Schneider, M., Beeres, K., Coban, L., Merz, S., Schmidt, S. S., Stricker, J., & De Smedt, B. (2017). Associations of non-symbolic and symbolic numerical magnitude processing with mathematical competence: A meta-analysis. *Developmental Science, 20*(3), p. 16.

Schoenfeld, A. H. (2007). Method. In F. K. Lester (Ed.), *Second handbook of research on mathematics teaching and learning* (pp. 69–107). Charlotte: Information Age Publishing.

Siegler, R. S., & Booth, J. L. (2004). Development of numerical estimation in young children. *Child Development, 75*, 428–444.

Siegler, R. S., & Lemaire, P. (1997). Older and younger adults' strategy choices in multiplication: Testing predictions of ASCM using the choice/no-choice method. *Journal of Experimental Psychology: General, 126*, 71–92.

Sinclair, N., & Jackiw, N. (2011). *TouchCounts*. Application for the iPad. https://itunes.apple.com/ca/app/touchcounts/id897302197?mt=8??

Sinclair, N., & Pimm, D. (2015). Whatever be their number: Counting on the visible, the audible, and the tangible. In M. Meletiou-Mavrotheris, K. Mavrou, & E. Paparistodemou (Eds.),

Integrating touch-enabled and mobile devices into contemporary mathematics education (pp. 50–80). Hershey: IGI Global.

Skwarchuk, S. L., Sowinski, C., & LeFevre, J. A. (2014). Formal and informal home learning activities in relation to children's early numeracy and literacy skills: The development of a home numeracy model. *Journal of Experimental Child Psychology, 121*, 63–84.

Stokes, D. E. (1997). *Pasteur's quadrant: Basic science and technical innovation*. Washington, DC: Brookings.

Watson, A., & Ohtani, M. (2015). *Task design in mathematics education: The 22nd ICMI Study*. New York: Springer.

Cited papers from Sun, X., Kaur, B., & Novotna, J. (Eds.). (2015). Conference proceedings of the ICMI study 23: Primary mathematics study on whole numbers. Retrieved February 10, 2016, from www.umac.mo/fed/ICMI23/doc/Proceedings_ICMI_STUDY_23_final.pdf

Baccaglini-Frank, A. (2015). Preventing low achievement in arithmetic through the didactical materials of the PerContare project (pp. 169–176).

Butterworth, B. (2015). Low numeracy: From brain to education (pp. 21–33).

Elia, I., & van den Heuvel-Panhuizen, M. (2015). Mapping kindergartners' number competence (pp. 177–185).

Gervasoni, A., & Parish, L. (2015). Insights and implications about the whole number knowledge of grade 1 to grade 4 children (pp. 440–447).

Gould, P. (2015). Recalling a number line to identify numerals (pp. 186–193).

He, S. (2015). How do Chinese students solve addition/subtraction problems: A review of cognitive strategy (pp. 194–202).

Ma, Y., Xie, S., & Wang, Y. (2015). Analysis of students' systematic errors and teaching strategies for 3-digit multiplication (pp. 203–211).

Milinković, J. (2015). Counting strategies and system of natural number representations in young children (pp. 212–219).

Mulligan, J., & Woolcott, G. (2015). What lies beneath? Conceptual connectivity underlying whole number arithmetic (pp. 220–228).

Obersteiner, A., Moll, G., Reiss, K., & Pant, H. A. (2015). Whole number arithmetic: Competency models and individual development (pp. 245–250).

Roberts, N. (2015). Interpreting children's representations of whole number additive relations in the early grades (pp. 245–250).

Sayers, J., & Andrews, P. (2015). Foundational number sense: The basis for whole number arithmetic competence (pp. 124–131).

Sinclair, N., & Coles, A. (2015). 'A trillion is after one hundred': Early number and the development of symbolic awareness (pp. 251–259).

Verschaffel, L., Torbeyns, J., Peters, G., De Smedt, B., & Ghesquière, P. (2015). Analysing subtraction-by-addition in the number domain 20–100 by means of verbal protocol vs reaction time data (pp. 260–267).

Yang, D-C. (2015). Performance of fourth graders in judging reasonableness of computational results for whole numbers (pp. 268–276).

Young-Loveridge, J., & Bicknell, B. (2015). Using multiplication and division contexts to build place-value understanding (pp. 379–387).

7

Artefacts and Tasks in the Mathematical Preparation of Teachers of Elementary Arithmetic from a Mathematician's Perspective

Bernard R. Hodgson

Introduction

The main focus of this paper is the mathematical preparation of primary school teachers in relation to the teaching of elementary – whole number – arithmetic. My comments are based mostly on my experience and reflections as a mathematician involved in the education of pre-service teachers, but with occasional inspiration from development activities for in-service teachers with whom I collaborated.

As a consequence, the prime intent of the paper is not to examine what may be happening about the learning of arithmetic by actual pupils in classrooms but rather to concentrate on the kind of 'adult experiences in mathematics' that, in my opinion, prospective teachers ought to encounter in order to better prepare for their role as guides accompanying their pupils in the acquisition of concepts and skills related to whole number arithmetic.

The context that induced the department of mathematics at my university to create two mathematics courses specifically for future primary school teachers – one of them being devoted to arithmetic – is briefly described in Hodgson and Lajoie (2015). Suffice it here to recall that this involvement has been ongoing now for more than four decades and that the responsibility of preparing prospective teachers for their mathematical duties is shared, and it goes without saying, with the Faculty of Education, where student teachers take three mathematics education courses (*didactique des mathématiques*, in French). The classroom reality and the pupils' needs are of course more significantly integrated in this didactical environment. An underlying theme of Hodgson and Lajoie (2015) is to stress the complementary roles played by mathematicians and mathematics educators (*didacticiens*) in this endeavour. It should however be noted, as indicated in the survey by Bednarz (2012, Tables

B.R. Hodgson (✉)
Université Laval, Quebec City, QC, Canada
e-mail: Bernard.Hodgson@mat.ulaval.ca

1 and 2), that the model adopted at my university in that connection is rather unusual in the Canadian context (including in the province of Québec).

In order to provide an insight into the arithmetic course (entitled *Arithmétique pour l'enseignement au préscolaire/primaire*) that we have developed for primary school teachers and crystallise its main intents, I wish in this paper to examine both its spirit and some of its components. This will bring me to consider the two central themes at the heart of Chap.(*Aspects that Affect Whole Number Learning: Cultural Artefacts and Mathematical Tasks*): the importance and variety of artefacts (often of a cultural and historical nature) that can be used to support the learning of whole number arithmetic, as well as the role played by mathematical tasks proposed in order to foster the 'mathematical message' that may be conveyed through the artefacts. I will discuss here concrete examples taken from our arithmetic course, intending by so doing to illustrate a crucial observation strongly emphasised in Chapter, namely that artefacts and tasks form an inseparable pair. However, I will first present some general observations about the mathematical preparation of primary school teachers with regard to whole number arithmetic.

Preparing Mathematically for the Teaching of Arithmetic

> One should not […] delay too late the moment when abstraction shall become the form and the condition of the whole teaching: finding for each pupil and for each study the right moment when it is advisable to move from the intuitive form to the abstract form is the great art of a true educator.[1] (Buisson 1911)

The philosophy underlying our arithmetic course is based on the conviction that in order to adequately fulfil their role as guides and become efficient communicators, primary school teachers should have developed such a level of mathematical competency that they see themselves as being in full possession of the mathematical tool with which they will be working, in other words that they feel *autonomous* with respect to their mathematical judgements concerning primary school arithmetic. We thus offer to the student teachers an opportunity for a personal reconstruction of elementary arithmetic through a mathematical pathway intended to allow them to clarify and develop basic notions underlying the teaching and learning of arithmetic at the primary level. This enterprise can be interpreted as aiming to both demystify and demythicise mathematics in general, but especially arithmetic, for prospective teachers.

[1] Original French text: 'Il ne faut (…) pas reculer trop tard le moment où l'on fera de l'abstraction la forme et la condition de tout l'enseignement: trouver pour chaque élève et pour chaque étude le moment précis où il convient de passer de la forme intuitive à la forme abstraite est le grand art d'un véritable éducateur'.

Hopefully, such an experience will allow them to gain confidence in their own expertise – of a very specific nature – about mathematics as seen in relation to the education of primary school pupils.

Part of what we aim at achieving on the subject of arithmetic with our student teachers is well captured by a quotation from the illustrious mathematician Leonhard Euler, the author, it should be reminded, of many influential textbooks intended for students. In the preface of an arithmetic textbook for Russian schoolchildren published in 1738 under the auspices of the St Petersburg Academy of Sciences, *Einleitung zur Rechenkunst* (*The Art of Reckoning*), Euler wrote:

> The learning of the art of reckoning without some basic principles is sufficient neither for solving all cases that may occur nor to sharpen the mind, as should be our specific aim. [...] Thus when one not only grasps the rules [of reckoning], but also clearly understands their causes and origins, then one will to some extent be enabled to invent new rules of one's own, and to use these to solve such problems, for which the usual rules would not be sufficient. One should not fear that the learning of arithmetic might thus become more difficult and require more time than when the raw rules are presented without any explanation. Because any individual understands and retains much more easily those matters, whose causes and origins he clearly comprehends.[2] (Euler 1738, pp. 3–4)

It is far from me to suggest that the aims or means of our arithmetic course are in any way new or revolutionary. For a very long while, a number of authors have been reflecting on the need to improve the mathematical preparation of pre-service school teachers and proposing varied and at times highly innovative approaches. One outstanding example is given by Felix Klein, the very first President of ICMI (1908–1920), who presented in the early twentieth century a famous series of lectures intended for teachers (Klein 1932). Although Klein was then mainly addressing secondary school teachers of mathematics, parts of his comments, especially in the very first chapter devoted to 'calculating with natural numbers', can be seen as pertaining directly to elementary arithmetic, and the needed mathematical background and vision with which primary school teachers, in his opinion, should be familiar. Another example, having as a setting my own university but at the time of the 'New Math' era, is given by Wittenberg, Sister Sainte Jeanne de France and Lemay (Wittenberg et al. 1963) – the three authors were then all connected to the mathematics department. Resisting the 'bourbakised' vision of mathematics teaching (p. 91) then quite fashionable, the authors reflect on the numerous reform movements of those days which, they claim:

---

[2]Original German text: 'Die Erlernung der Rechenkunst ohne einigen Grund weder hinreichend ist, alle vorkommenden Fälle aufzulösen, noch den Verstand schärfet, als dahin die Absicht insonderheit gehen sollte. [...] Dann wann man auf diese Art nicht nur die Regeln begreift, sondern auch den Grund und Ursprung derselben deutlich einsieht, so wird man einigermassen in Stand gesetzt, selbsten neue Regeln zu erfinden und vermittelst derselben solche Aufgaben aufzulösen, zu welchen die sonst gewöhnlichen Regeln nicht hinreichend sind. Man hat auch im geringsten nicht zu befürchten, dass die Erlernung der Arithmetik auf diese Art schwerer fallen und mehr Zeit erfordern werde, als wann man nur die blossen Regeln ohne einigen Grund vorträgt. Dann ein jeder Mensch begreift und behält dasjenige im Gedächtnis viel leichter, wovon er den Grund und Ursprung deutlich einsieht'.

> at times seem to lock themselves into a surprising and naïve conviction that it is enough, in that domain, to meditate on the thinking of a single man (multiplicatively reincarnated, it is true), and that: *whoever has read Bourbaki, has read everything*.[3] (Wittenberg et al., p. 11)

The authors propose a 'genetic approach' as a way to allow practising and prospective teachers to see elementary mathematics with new eyes – so to say, to see it as their pupils will – and to reflect on its internal structure (p. 13).

We do not aim, in our arithmetic course for teachers, at presenting a fully fledged 'genesis' of the basic concepts related to whole numbers. Still we wish to adhere to a vision that remains as primitive as possible. For that purpose our arithmetical journey relies, to a large extent, on a rudimentary yet very fruitful numerical artefact for tackling numbers: sequences of tallies (Section). This allows us to gradually build a body of knowledge about the (set of) whole numbers, a patent emphasis being put on establishing, with a certain level of rigour, the basic properties at stake. It is thus a structural perspective on elementary arithmetic that we propose to our student teachers, which brings into the discussion a level of abstraction possibly rather new, maybe even strange, to some of them. But as expressed by the renowned French educator Ferdinand Buisson at the turn of the twentieth century (see the quotation as epigraph to this section), such an abstract perspective is at the heart of the teaching and learning process.

Our mathematical approach to 'arithmetic for teachers' has clear links to several current or recent research works, such as that of Grossman, Wilson and Shulman (1989), stressing the importance of a teachers' sound knowledge of mathematical content. It also has connections with the famous study of Ma (1999) concerning a 'profound understanding of mathematics', as well as with the work of Ball and Bass (2003) about 'mathematical knowledge for teaching'. In that context, I think it is of interest to comment, from a mathematician's perspective, on some of the mathematical artefacts and tasks used in our teaching of arithmetic.

Accesses to the Concept of Number

'How do numbers emerge?' Such is the question raised by Hans Freudenthal, the eighth President of ICMI (1967–1970), at the very outset of a chapter entitled 'The Number Concept: Objective Accesses' from his monumental *Mathematics as an Educational Task* (1973, p. 170). He proposes a fourfold distinction, discussing successively – from a mixture of mathematical and didactical vantage points – the emergence of the number concept under the disguises of *counting* number, *numerosity* number, *measuring* number and *reckoning* number.

---

[3]Original French text: 'alors que prolifèrent des entreprises de réforme [qui] semblent parfois s'enfermer dans une surprenante et naïve conviction que c'est assez, dans ce domaine, de méditer la pensée d'un seul homme (multiplicativement réincarné, il est vrai), et que: *qui a lu Bourbaki, a tout lu*'.

(Another way of reacting to the question put forward by Freudenthal is found under the heading 'The Logical Foundations of Operations with Integers' of Klein (1932), notably pp. 11–13. While Klein sees issues of psychology and epistemology being at stake, he chooses, as indicated by the title of that section, to propose a mathematical reflection based on a spectrum of perspectives and arguments from logic, including a purely formal theory of numbers.)

Space prevents me here from entering into all the nuances of Freudenthal's very rich discussion of the various accesses he discusses in relation to the concept of number. But I would like to use his framework as a basis for my reflections and for establishing connections to some of the core ingredients of our arithmetic course for primary school teachers.

Counting Number

Freudenthal characterises the notion of counting number as being connected to 'the reeling off in time of the sequence of natural numbers' (1973, p. 170). He observes that grasping 'the whole unlimitedly continuing sequence' of numbers is for children 'a conceptual seizure that has no analogue in learning the names of colours and letters' (1973, pp. 170–171). At stake here is the notion of *successor*, at the heart of the axiomatic approach to arithmetic proposed by Giuseppe Peano, jointly with the principle of mathematical induction. As a consequence of the successor is de facto an order implicitly introduced: we are thus in presence of the general notion of ordinal number.

One possible implementation of the idea of counting numbers is via an artefact that actually plays a fundamental role in our course, sequences of tallies. The idea of successor is then readily perceived as the simple adjunction of a new tally to a given sequence, and this can 'obviously' be repeated indefinitely, at least in principle. I shall return to this particular artefact in Section.

One very early encounter of children with counting numbers is when learning a counting list, the successive terms being either written as numerals or expressed as words in their mother tongue. The memorising of the usual oral counting list is often supported by nursery rhymes. A possible task that may be proposed to teachers in that connection is to examine if an actual poem or a song could really be used for counting. What are the qualities that a good 'counting song' (or list) should possess? How far could one go with a particular song like *Au clair de la lune* (see https://en.wikipedia.org/wiki/Au_clair_de_la_lune) – for instance, in order to count the number of pupils in the classroom? (Other aspects of songs and poems being used in arithmetic are discussed in Section.)

One particular task I like to give to my student teachers on the very first day of the course is to build a written counting list using the symbols from some given alphabet. (It is understood here that the alphabet comes with a specific order among the symbols it contains.) Without having yet discussed the notion of positional-value numeration system, I propose to restrict the available digits to, say, 0, 1, 2 and

3 (in that order), and invite them to build a counting list, in the same spirit as our usual list of numerals, but using only these symbols. While some may come back with rather original lists, most of them would have built a list 'in base four', analogous to the usual base ten list. But given the alphabet comprising the symbols A, B, C and D, the typical answer

A, B, C, D, AA, AB, AC, AD, BA, BB, …, DD, AAA, AAB, …

clearly illustrates that the usual counting list they know very well (base ten) brings into play one symbol having a rather special behaviour: 0.

The learning of the usual oral counting list may introduce some linguistic peculiarities, usually specific to a given language. For instance, an interesting cultural task (in French) is to observe the distinction, when counting by tens, between the 'regular'

cinquante, soixante, septante, octante, nonante

and the more usual (but depending on countries)

cinquante, soixante, soixante-dix, quatre-vingts, quatre-vingt-dix

the latter being (partly) a remnant of vigesimal numeration (see this volume, Sect.[4]

Numerosity Number

Using even animals as *cas de figure* for small numerosities, Freudenthal notes that 'perhaps the numerosity number is genetically earlier than the counting number' (1973, p. 171). His comments point to the fact that recognising at a glance – without counting – the number corresponding, say, to four dots (even if placed randomly), is an easy task (this capacity, called *subitising*, is described in this volume, Sect.): one can instantly 'see' the four dots. But the same is possibly not true for most people when looking quickly at the dots in Fig. a. However, the pattern used in Fig. b is such that the numerosity of the dots is immediate. (See also Sect. for comments about strategies related to artefacts with a structural feature such as that of Fig. b.)

Numerosity rests on the possibility of identifying the 'number' corresponding to a certain situation without numbering the objects one by one. The idea is to associate the given situation with another one, the question at stake then being not 'how many?' but rather 'is it as many as?'. While the notion of equipotency (or one-to-one correspondence) on which rests this approach to whole numbers is a very natural one (and it will play a central role when discussing sequences of tallies in Sect.),

[4]One may be reminded here, for instance, of Molière's *L'Avare* (1668), when one of Harpagon's servants, fawning over him about his longevity, says: 'Par ma foi, je disais cent ans, mais vous passerez les six-vingts' (Act II, Scene 5).

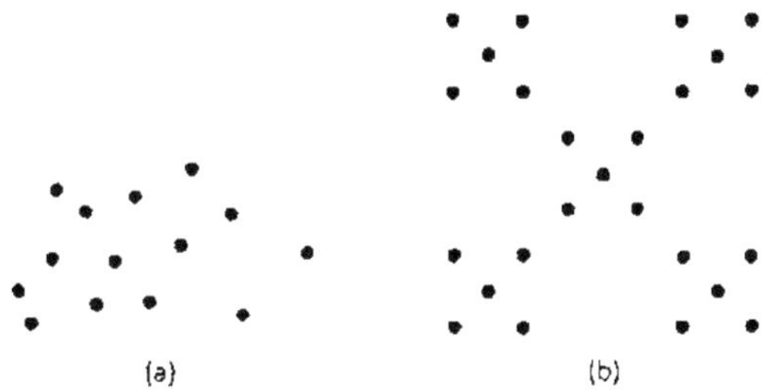

Fig. Numerosity of dots aggregates (**a**): Randomly displayed dots (**b**): A pattern of dots

its formalisation via the general notion of cardinal number *à la Cantor* is definitely questionable, according to Freudenthal. (It may be noted here that Klein (1932, p. 12) speaks on the contrary in enthusiastic terms of this 'modern' approach due to Cantor.) The criticism made by Freudenthal (1973, p. 181) is as strong as can be:

(1) The opinion that the numerosity number, that is the potency, suffices as a foundation of natural numbers is mathematically wrong.
(2) The numerosity aspect of natural numbers is irrelevant if compared with the counting aspect.
(3) The numerosity aspect is insufficient for the didactics of natural numbers.

Freudenthal then spends more than fifteen pages expounding his objections.

Hodgson and Lajoie (2015, p. 309) mention that the first versions of our arithmetic course, in the 1970s, used a set-theoretic context to introduce natural numbers as cardinalities of finite sets, operations on numbers being defined via set-theoretic operations. This was in accordance with the spirit of the times, as can be seen through a quotation from a report of an ICMI-supported workshop organised by UNESCO in 1971, in a chapter on 'Primary Mathematics':

> All modern reformed programs have introduced the study of sets into mathematical instruction. This topic is perhaps the most visible trait of an actual change in primary mathematics teaching. [...] There is a universal trend to use sets to develop the concept of cardinal or natural numbers, and the four rational operations on natural numbers. (UNESCO 1973, pp. 5–6)

It was eventually decided to approach whole numbers in our course not on the basis of a possible prior idea of sets, but more intrinsically as 'primitive' objects of their own, as well as the operations defined on them: sequences of tallies thus entered the picture.

Measuring Number

Freudenthal discusses the notion of measuring number in a general context, comparison with a given unit leading sometimes to exhaustion of the magnitude to be measured and sometimes to incomplete exhaustion. The latter case can be seen as giving rise either to division with remainder (most appropriately called Euclidean division) or, when the unit is divided, to fractions. Eventually issues of commensurability and incommensurability, in an ancient Greek spirit, may come into play.

Fig. The factors of 18, *à la Cuisenaire*

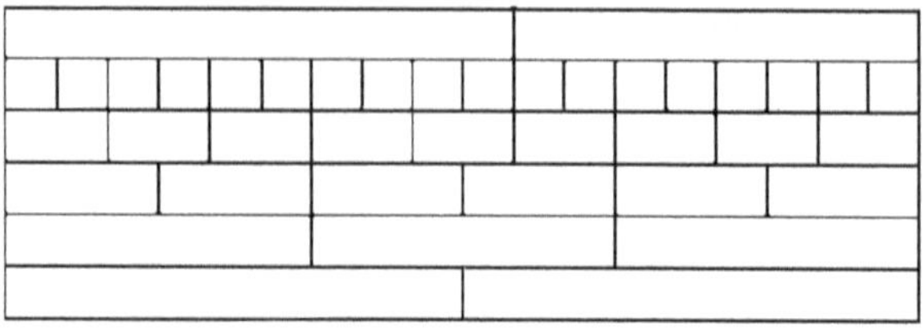

Central notions of elementary theory of numbers (the αριθμητική *arithmētikē* of the Greece of Antiquity) can be brought to the fore via the notion of measure: this is precisely how divisibility is introduced in Euclid's *Elements*. Two most fruitful artefacts to be used here are Cuisenaire rods.

Although strongly attached, in the collective memory, to the 'New Math' era, the Cuisenaire rods have a strong merit of their own, and it may be seen as still pertinent for future teachers to be familiar with this artefact. It is thus a pity that many of our student teachers have never encountered them earlier in their own schooling (as if legions of boxes of Cuisenaire rods were left sleeping in school basements). The simple fact of observing with the rods how a 'train' of length, say, 18 (one orange rod and one brown rod), can be divided into series of identical 'wagons' in five different ways (eighteen white, nine red, six light green, three dark green, two blue) is without any doubt a most inspiring way of addressing the notion of factors (see Fig.).

A most relevant distinction can be made here, considering, for instance, the factorisation $18 = 2 \times 9$, between the 'unit' 2 being repeated nine times and the 'unit' 9 repeated twice.

Transfer, in more generality, to the number line is then immediate. As stressed in Sect., line segments are precisely how Euclid 'saw' numbers, including whole numbers. And the issue of a number 'measuring' or not another one can readily be addressed via segments on the number line.

I wish to stress here another fertile artefact from ancient Greece for approaching elementary theory of numbers: the use of figurate numbers, that is, of certain geometrical arrangements of collections of dots. This vision of whole numbers does not emphasise an idea of measure, as I just discussed. Going back to the Pythagorean school, it provides a dynamic context revealing rich relations between given family of numbers. For instance, passing from a given square of side n to a square of side $n + 1$ involves the adjunction of a 'gnomon' (γνώμων) made of twice the length n and a unit. Fig. a presents the situation, for the case $n = 4$, in a figurate number style, while Fig. 10.3b uses a more traditional image based on area. It may be pointed out here that the latter artefact has as well a very long history, being present in essentially all the ancient mathematical traditions. Both artefacts of Fig. can serve as supports for *visual proofs* – in the present case, of the fact that $(n + 1)^2 = n^2 + (2n + 1)$.

Fig. Passing from the square of a whole number to the next

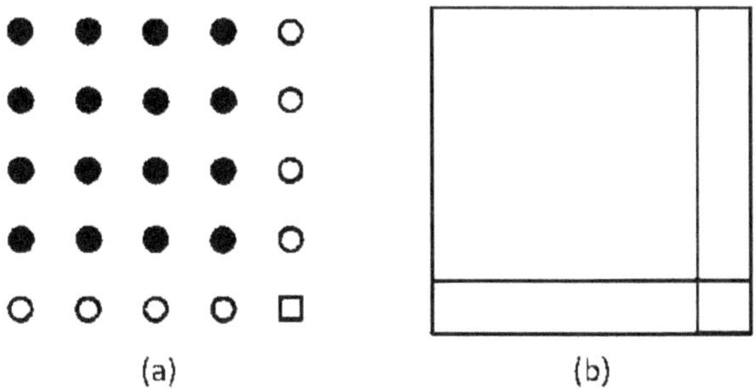

$$
\begin{aligned}
23 \times 15 &= (2 \times 10 + 3) \times (1 \times 10 + 5) && (1)\\
&= ((2 \times 10 + 3) \times (1 \times 10)) + ((2 \times 10 + 3) \times 5) && (2)\\
&= ((2 \times 10) \times (1 \times 10) + 3 \times (1 \times 10)) + && (3)\\
&\qquad ((2 \times 10) \times 5 + 3 \times 5) &&\\
&= ((2 \times 10) \times 10 + 3 \times 10) + ((2 \times 10) \times 5 + 3 \times 5) && (4)\\
&= ((2 \times 10) \times 10 + 3 \times 10) + (5 \times (2 \times 10) + 3 \times 5) && (5)\\
&= (2 \times 10 \times 10 + 3 \times 10) + (5 \times (2 \times 10) + 3 \times 5) && (6)\\
&= (2 \times 10 \times 10 + 3 \times 10) + ((5 \times 2) \times 10 + 3 \times 5) && (7)\\
&= (2 \times 10 \times 10 + 3 \times 10) + (10 \times 10 + 15) && (8)\\
&= (2 \times 10^2 + 3 \times 10) + (10^2 + 15) && (9)\\
&= (2 \times 10^2 + 3 \times 10) + (1 \times 10^2 + 15) && (10)
\end{aligned}
$$

Fig. Fragment of the calculation of 23 x 15 (detailed style)

Reckoning Number

Freudenthal (1973, p. 171) uses the expression reckoning number in order to highlight the algorithmic aspect attached to whole number arithmetic. This is the *logistikē* (λογιστική) of ancient Greece, that is, actual calculations involving operations of elementary arithmetic and, in particular, of course, the 'four operations'. We are now in the same ballpark as with Euler's *Rechenkunst* mentioned above.

Reckoning is considered from two different perspectives in our arithmetic course for teachers. One encounter with reckoning happens after about a month of the course, once the basic laws of arithmetic have been introduced and that positional-value numeration (base ten) has been fully reviewed, so that it is then possible to discuss and justify in a thorough manner the functioning of algorithms. A specific task proposed to teachers, concerning standard algorithms, is to identify minutely the way the arithmetic laws enter into action in a given algorithm. For example, concerning the multiplication 23 x 15, we provide them with a truly detailed com-putation (see Fig.) where the first ten lines from a series of twenty one are reproduced. A single mathematical basic event is enacted on each of these lines, and the students' task is to identify it.

Of course, we do not ask our students to construct by themselves calculations of such a type, as this is not really a task on which we would want them to devote energy and time. Nor do we have in mind them later using such a level of scrutiny with their own pupils. But for a teacher, seeing precisely (at least once in one's life-time) the way the basic arithmetic laws intervene in standard algorithms is, we believe, a most valuable experience.

Another task part of the same discussion is to have our students look for, look at and understand non-standard algorithms for the four operations. We also present them other artefacts of a historical flavour, such as, for multiplication, the gelosia method or Napier's bones the Egyptian algorithm.[5]

The other perspective we introduce in our course concerning reckoning happens on the very first day of the course: we give them as a task (to be done before the next class) to use their knowledge of algorithms for the four operations – these algorithms then play the role of artefacts – so to be able to compute in bases other than ten. We launch the task with them, reminding very briefly the idea of base ten numeration and asking then what if we were using, say, base eight. We make sure that they will go back home on that day having a reasonable intuition of what it means for a numeral to be expressed in a non-usual base such as eight, so that 'all that remains' for them is to do the calculations, using the knowledge they already have of basic arithmetic algorithms. Along the way we make them aware that in order to play the game fully in base eight (and not to 'cheat' via base ten), they will need to have access to information about additions and multiplications of one-digit numbers. An extra implicit task is thus for them to construct by themselves the artefacts for calculations in this new environment, namely, the Pythagorean tables for addition and multiplication in base eight. (See Section for comments on Pythagoras tables from the perspective of an artefact.) The next class rests on the work they will have done in the meanwhile.

We eventually tell them explicitly that an aim behind this task is to destabilise them to a certain extent with regard to basic arithmetic skills that may seem, at first sight, trivial. This way a context is created enticing them to get into a deep reflective mode about algorithms that they already know how to perform but are probably not in a position to explain or justify. Many students later testify to this 'non-base-ten calculation' task as a genuine thought-provoking moment for them.

Defining and Representing Whole Numbers

In passages about basic arithmetic laws in the literature for teachers, one may find comments linking, say, the commutativity of addition to expressions such as

$$345 + 67 = 67 + 345.$$

While this equality may be seen as a fine illustration of the property, the use of such a vision introduces a confusion between many aspects related to numbers, in particular between the nature or essence of natural numbers (or of the operations

---

[5] https://en.wikipedia.org/wiki/Ancient_Egyptian_multiplication

defined on them) and the representation of these numbers via a numeration system – however important the latter may be in practice.

It is for us of crucial importance in our course for teachers to introduce the natural numbers ‘in themselves’, without any reference to a system for representing them. In the early days of the course, they were introduced as cardinalities of (finite) sets. A major shift occurred in our approach to basic arithmetic when it was decided to restrict sets to a role of ‘linguistic’ tools for communication, instead of primitive concepts on which the whole arithmetical building should be based (see Section above) and to use the (historically primitive) notion of a *tally* (see this volume, Section) to introduce whole numbers.

Tallies: A Fruitful Artefact for Whole Numbers

> All writing may be put down, and nothing used but the score and the tally. Shakespeare (1594), *Henry the Sixth*, Part 2, Act IV, Scene vii (OED 2016)

Natural numbers, it was mentioned earlier, can be captured in a robust way by thinking of them as *counting* numbers (see Freudenthal’s comments in Section). In a written form, this vision can be rendered concretely through a basic artefact, the notion of *tally*,[6] as well as *sequences of tallies*.

A natural number is ‘naturally’ defined as a sequence of tallies – a *finite* sequence, of course. Accepting such a sequence to be eventually empty is not a major issue (especially with adults) and allows the introduction of 0, of fundamental importance when addition enters into the picture. The set of natural numbers is thus constituted of all the finite sequences of tallies, and this can be accepted as a working definition with prospective teachers.

Leaving aside for the moment the empty sequence of tallies (a special symbol, such as an inverted triangle, could be introduced for that purpose), the (unlimited) sequence of counting numbers thus begins

$$| \qquad || \qquad ||| \qquad |||| \qquad ||||| \qquad |||||| \qquad \cdots$$

This artefact explicitly allows the notion of successor to be fully seen in action.

In order to gain a necessary level of generality, a notation such as

$$\overbrace{||\cdots||}^{n}$$

may be introduced in order to represent a sequence of tallies of arbitrary length, n.

[6] Other words traditionally used in a same sense include *notch, score* or *stroke*. A tally is to be considered simply as a mark, typically a short line segment. In our course, we speak of the ‘*bâton*’ (in French), i.e. the *stick*.

Once we have such concrete models for natural numbers, a fundamental notion to discuss is the equality of two given natural numbers, which is to be captured through the verification that the corresponding sequences of tallies are identical. The natural way to render this idea is via the establishment of a bijective link between the two sequences. In the present context, this scheme of one-to-one correspondence between sequences of tallies appears as a most natural artefact, requiring no sophisticated set-theoretic support. It also leads to the definition of order among natural numbers, when one sequence happens to be exhausted before the other in a search for a one-to-one correspondence.

Such a setting in turn allows for operations on natural numbers to be introduced through operations on sequences of tallies which, in that context, can be accepted as natural and primitive. For instance, the addition of two given numbers n and m is defined as the juxtaposition of the corresponding sequences of tallies:

$$\overbrace{||\cdots||}^{n+m} \;=\; \overbrace{||\cdots||}^{n}\;\overbrace{||\cdots||}^{m}$$

(with equality here being per definition). The sum $n + m$ is readily seen to be a natural number. In a similar vein, the multiplication of n and m can be defined as the result of replacing each tally in the sequence for n by a replica of the sequence for m. For matters of convenience, the natural number n x m obtained as the product sequence can be displayed as a rectangular array (or matrix) of tallies:

$$n\left|\begin{array}{c}\overbrace{||\cdots||}^{m}\\ ||\cdots||\\ \vdots\\ ||\cdots||\end{array}\right.$$

From these definitions (using the notion of the empty sequence of tallies) follow immediately, for instance, two basic arithmetical facts: when a sum $n + m$ is 0, then both terms are 0; and when a product n x m is 0, then at least one of the factors is 0.

It may be noted that a similar artefact for counting numbers is the model of boxes of aligned dots used by Courant and Robbins (1947, p. 2 et seq.) in their study of the laws governing the arithmetic of whole numbers. Addition then corresponds to 'placing the corresponding boxes end to end and removing the partition' (p. 3):

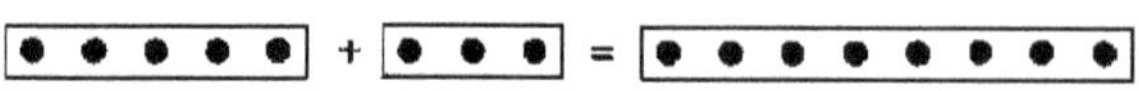

while the multiplication n x m is defined via a box with n rows and m columns of dots (eventually reorganised as a box of aligned dots).

Establishing the Basic Laws of Arithmetic

The tally artefact has as a bonus the feature that the basic properties of arithmetic can be actually *proved* and not merely stated or illustrated. For instance, the property of commutativity of addition, mentioned at the beginning of this section, then amounts to the following: given two arbitrary sequences of n and m tallies, one shows that the order of juxtaposition does not matter by an appeal to the obvious one-to-one correspondence – the rightmost tally of m in the sequence $n + m$ is linked to the leftmost tally of m in $m + n$ and so forth. Both sequences $n + m$ and $m + n$ will then be exhausted simultaneously.

In turn all other fundamental properties of addition and multiplication can be proved similarly, thus leading to the establishment of 'the fundamental laws of reckoning' (associativity and commutativity of + and x, identity elements, compatibility of = with + and x, simplification for + and x, distributivity x/+, laws concerning order) (see, e.g. Klein (1932), p. 8, where such rules are simply stated). It is important, with teachers, to stress that these properties speak of the behaviour per se of numbers and certain operations defined on them and, as such, are totally independent of numeration, i.e. of ways of 'writing down' the numbers in a given system.

It may be noted that the artefact of tallies, as a concrete model for whole numbers, comes with practical limitations: for instance, given two sequences made, respectively, of, say, one thousand and one thousand and one tallies, it would be quite cumbersome to compare their size by searching for a possible one-to-one correspondence. But such is not the main point behind this artefact: the really crucial aspect is that we have a precise agreement about what any given natural number 'is'. Such a vision, independent of any numeration system, is a fundamental awareness for a teacher.

The interested reader will have noticed that we are now very close to the principle of mathematical induction and the full set of Peano axioms. While mentioning this en passant, we do not see it as appropriate to insist on such an abstract vision in our course for primary school teachers.

Representing Whole Numbers

Of course, the time comes, in our arithmetic course for teachers, when the concept of number emerging from the preceding tally-wise definition should be related to standard arithmetic practice, and in particular to our usual positional-value numeration system in base ten. The many devices and variants for representing numbers developed throughout history turn out to be very informative artefacts in order to distinguish between two fundamental issues at stake when writing or recording numbers: establishing a numeration scheme for memory or communication purposes, and establishing one for reckoning purposes.

Fig. The 'cargo method' of F. Lemay

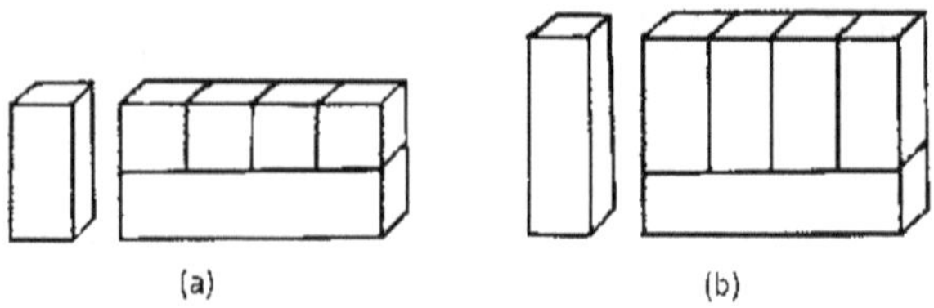

Many interesting artefacts are available, often of a strongly cultural nature, as regards memory or communication devices for numbers: tallies, Egyptian numeration, Roman numeration, quipus, etc. When emphasis is put on devices facilitating calculations, a different set of artefacts may come to mind – in particular abaci – where the notion of position is part of the artefact itself. (See Sect.for a rich catalogue of artefacts for whole number arithmetic used throughout history.)

The power of the positional nature of our usual numeration system should not be underestimated when it comes to consider it from a reckoning perspective.[7] Devices emphasising position thus play a crucial role in that connection. Many such artefacts are discussed in Chapter, in particular in Sects.(multibase arithmetic blocks) and (spike abaci). In the case of the blocks, the position is transmitted via the size of blocks, while for the abaci it is the place of a given spike on the physical device that indicates position.

Mention is made in Sect. of the possibility of using beads of different colours being piled on the various spikes of an abacus in order to make more evident the difference in the roles played by the beads. The authors then observe that such a use of colours 'seems not advisable, as attention is focused on colours and exchange conventions rather than on order and position'.

While agreeing fully with the importance of stressing order and position when learning about positional value, I would suggest that an artefact integrating colours, and even emphasising exchanges through a colour code, may be of interest on its own, at least when working with teachers.

I wish to briefly describe here an artefact, due to Lemay (1975), based on the Cuisenaire rods and using both the lengths and the colours of the rods in order to develop the core for a positional-value system. The so-called 'cargo method' of Lemay (*méthode des cargaisons*, in French) leads to a mechanical way for exchanging, without counting, a certain sets of rods for a rod of the 'next type'.

We first agree on a certain rod serving as a basis for numeration – the word basis being taken here as well in a truly physical sense. In the example of Fig., we use the pink rod (of length four) as the basis, lain horizontally. The exchange pro-cess rests on the following rule: if a number of rods of the same colour can be placed upright side by side so as to cover it completely, it can be exchanged for a rod whose length matches the total height of the freight (still standing vertically on the basis).

[7] One may be reminded here of a famous engraving from Gregor Reisch's *Margarita philosophica* (1503), where an allegorical figure of Arithmetic appears to express her preference in a kind of calculation contest between the 'Ancient' and the 'Modern' (see Swetz and Katz 2011).

Figure shows a cargo of white rods being exchanged for one red rod, and Fig., a cargo of red rods replaced by a light green one.

This method clearly leads to a fully fledged positional-value numeration system (base four in the example), the length of the successive rods corresponding to the position of digits in a given numeral.

Such an artefact could be related to one involving strictly a colour code, the Cuisenaire rods being simply replaced by colour tokens. The exchange rule then requires grouping the tokens of a given colour (e.g., by counting) so as to form a heap to be exchanged for one token of the next colour. Although possibly considered more abstract, as no physical indication for the 'weight' of a digit (i.e. its position) is conveyed by the colour itself, such a system appears of unquestionable importance. It corresponds, for instance, to two important historical artefacts, the (additive) Egyptian and Roman numeration systems. The fact, in the latter case, that the symbol L, say, corresponds to fifty has no (immediate) physical connection – although the history of the Roman characters may be of interest on its own, as plainly shown by Ifrah (2000, pp. 187–200).

It is thus of importance to convey the idea that the value of a given element may depend strictly on the agreement made about it and not on its physical size. While clearly of fundamental interest for positional numeration, multibase arithmetic blocks, for that matter, do not convey the whole story of numeration. Non-physical (or abstract, if one wishes) exchange codes are also present in concrete artefacts (and daily situations), such as monetary systems. Among the Canadian coins, for instance, the 10¢ coin is smaller than the 5¢ coin, but children have no problem agreeing to this arbitrary value. (The fact that the 'unit', namely, the 1¢ coin, has recently disappeared from physical monetary transactions in Canada raises other interesting numerational issues, as the 5¢ does not really correspond to a heap of five 1¢ but rather takes its value from an abstract agreement.)

Historical, Logical and Didactical Background to the Tallies

Hodgson and Lajoie (2015) offer comments on the long and diverse history of the use of tallies as an artefact for counting numbers, from 'early stone age' to more recent centuries. Using passages from Ifrah (2000), they recall how present this approach to numbers is among many cultures.

Sequences of tallies can also be seen as a practical artefact of ancient times, but one still largely in use even today when counting a not-too-large population, often implemented by means of groupings by five:

In a modern context, this 'unary' vision of natural numbers is often encountered in works pertaining to logic, be it in an epistemological context, as 'a primitive form of notation for natural numbers' (Steen 1972, p. 4), or in relation to the notion of computability as defined via Turing machines (Kleene 1952, p. 359) (Davis 1958, p. 9). It may be of interest to note in the latter case that in order to have access to a simple notation for 0, a sequence of $n + 1$ tallies is used to represent the natural number n as a 'tape expression' on the Turing machine. Steen emphasises the generative aspect of sequences of tallies, starting with the sequence made of a single tally and accepting as a construction rule the adjunction of a tally to a given sequence (see also Lorenzen (1955, p. 121 et seq.), under a chapter entitled 'Concrete mathematics'[8]).

As an example of a recent didactical application, this '*constructive* (or operative) *foundation* of natural number' presented by Lorenzen, and in particular the 'calculus of counting actions' just described, is acknowledged by Wittmann (1975, p. 60) as the basis of the reflections he proposes about the teaching and learning of natural numbers.

A Miscellany of Artefacts and Tasks for Elementary Arithmetic

I now mention briefly a few other artefacts and tasks pertinent to elementary arithmetic and used in our course.

An Artefact for Focusing on Remainders: Clock Arithmetic

Today is Tuesday. What day of the week will it be in 18 days from today? Or: *It is now 15:30. What time will it be in 1000 hours from now?* Such questions, connected to daily basic arithmetic, emphasise the fact that in many contexts related to division, the remainder may be seen as more relevant that the quotient. Being familiar with the 12-hour (and the 24-hour) clock is an important basic learning. And this in turn may be the starting point supporting the main ideas of modular arithmetic in general, using as an artefact so-called clock arithmetic.

---

[8] It may be noted that another artefact for natural numbers found in logic, of a more advanced nature, is provided by the so-called von Neumann ordinals (von Neumann 1923). They amount to have each ordinal be defined as the set of ordinals that precede it (the empty set being taken as the starting point, ordinal 0). In such a context, we have, for instance, $3 = \{0,1,2\}$, that is, 3 is a specific set with three elements. A nice feature of this definition of ordinal numbers is that it allows easily the transfer to transfinite ordinals. But we are then a bit beyond primary school arithmetic.

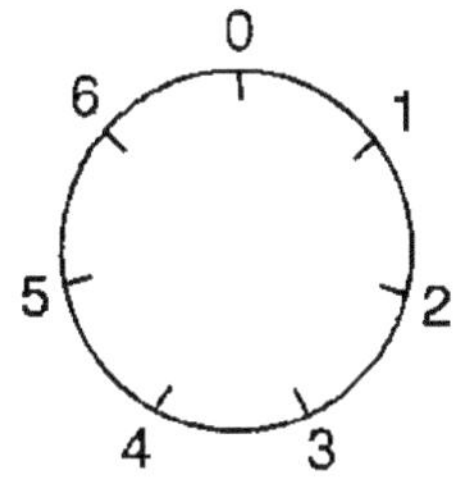

Fig. The 7-hour clock

Fig. The six-column Eratosthenes sieve (for primes up to 50)

2	3	4	5	6	7
8	9	10	11	12	13
14	15	16	17	18	19
20	21	22	23	24	25
26	27	28	29	30	31
32	33	34	35	36	37
38	39	40	41	42	43
44	45	46	47	48	49

Figure shows a slightly special clock, namely, a '7-hour' clock (with 0 used instead of 7, in distinction from a standard clock). It is easy to define basic arithmetical 'clock operations' in this environment, such as addition: 2 + 18 = 6 (which may be read as today being Tuesday, it will be Saturday in 18 days). Subtraction and multiplication are as well easily implemented: we are thus in the vicinity of the ring ***Z*/*nZ***.

An Artefact for Finding Prime Numbers: The Six-Column Sieve

The sieve of Eratosthenes is a well-known device for finding prime numbers. Experience shows that students will very frequently list the natural numbers in ten columns in order to do the sieving. However, using as an artefact a six-column sieve (Fig.) immediately reveals a nice phenomenon, once the proper multiples of the first two primes have been eliminated: *with the exception of* 2 *and* 3, *all prime numbers are of the form* $6k + 1$ *or* $6k - 1$. This is a fine example of a visual proof: the artefact in itself 'is' the proof of this result. The sieving process is then made considerably easier due to the format of the sieve. See Hodgson (2004, pp. 334–335) for further details.

A crucial related task here is of course to ask: why six columns? A bit of elementary theory of numbers – either using 6-hour clock arithmetic or examining a prime p jointly with its two neighbours, $p - 1$ and $p + 1$ – will be helpful in bringing out the fact that 'we knew' already that any prime, besides 2 and 3, is the neighbour of a multiple of 6. However, the artefact in itself is still of interest in primary education.

An Artefact for Observing Divisors: A Brick in the Wall

I have mentioned above (see Section) how Cuisenaire rods may be used for showing in action, so to say, the divisors of a given number. The same artefact may be used for finding concretely common divisors of two numbers, and this way 'see' the GCD. A similar remark applies to common multiples and LCM of numbers.

An Artefact for Applying Divisors: The Long Hotel

A well-known problem in the literature for primary school teachers is the 'Long Hotel' problem, using the parlance of Cassidy and Hodgson (1982). The underlying artefact is set as follows: there are n rooms along a long corridor, and the n guests consecutively apply an 'open/close' process on the doors, Guest #k changing the position of every k-th door (starting with door #k). The question of determining which doors are left open and which ones are closed at the end of the process boils down to identifying the divisors of a given door number, and more precisely the parity of the number of divisors: the perfect squares here stand out as of special interest.

An Artefact for Applying GCD and LCM: The Circle Hotel

Cassidy and Hodgson (1982) introduce a variant to the preceding problem, using the 'what-if-not' strategy in problem posing and transferring the process to a circular corridor. The 'Circle Hotel' problem provides a nice context to see clock arithmetic in action. It turns out in this case that a single door remains open at the end of the process, the precise nature of the door number depending on the parity of n.

Various additional questions can be asked in this context, related, for instance, to the number of times Guest #k will go round the corridor before stopping (i.e. before touching the same doors again) or the number of doors Guest #k will have touched during the process. The answers to these questions have to do with the GCD and the LCM of n and k and are connected to a famous artefact from a few decades ago, the so-called Spirograph. Figure shows the figure generated with the Spirograph by rolling a small wheel of 30 teeth inside a big wheel of 105 teeth, which corresponds to the action of Guest #30, when the Circle Hotel has 105 rooms.

Considered as an artefact, the Spirograph is of special interest on its own, as the elegance and beauty of the figures it can generate may play the role of 'attention-catcher' and invite to indulge in further investigation, for instance, about the family of star polygons $\{n/d\}$ (Fig. b). These are (generalised) polygons obtained by connecting with line segments every d-th point on a circle where n equally spaced points are marked. Such aspects are discussed in Hodgson (2004, pp. 324–328).

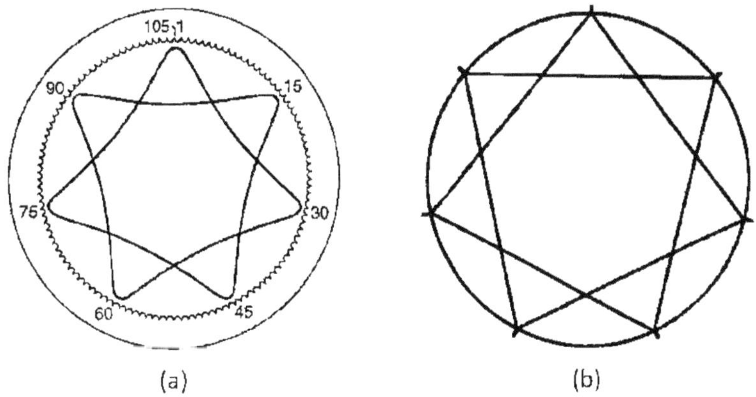

Fig. The Spirograph curve 105/30 and the star polygon {7/2}

An Algorithmic Artefact: The Euclidean Algorithm

Understanding and applying the Euclidean algorithm for finding the GCD of two numbers can be seen as a task appropriate for prospective primary school teachers. It has a very positive impact, as this algorithm is totally new to most student teachers. This algorithm provides the occasion for a nice learning experience, as prospective teachers are then in the same situation as their future pupils with respect to encountering a certain algorithm for the first time.

This algorithm, via Bézout's identity, then becomes a fine artefact for solving 'popular' problems, such as whether it is possible or not to obtain a certain quantity of water using two pails whose capacity is known.

Visual Artefacts: Proofs Without Words

The use of a figure for supporting the proof of – or even for 'proving' by itself a given arithmetic identity has a long history. Already in Euclid's *Elements,* for instance, are results with a substantial visual component (but such was not of course the spirit of Euclid's approach). A classical example is about the area of the square of side $a + b$ (proposition II.4, accompanied by a figure similar to Fig. b above). Another result of Euclid (proposition II.1) concerns the area of a large rectangle that has been divided into smaller rectangles (Fig. a). In modern terms, the situation can be interpreted as corresponding to the distributivity of multiplication over addition.

In a similar spirit, Klein (1932, p. 26) proposes Fig. b as a support for the formula:

$$(a-b)(c-d) = ac - ad - bc + bd.$$

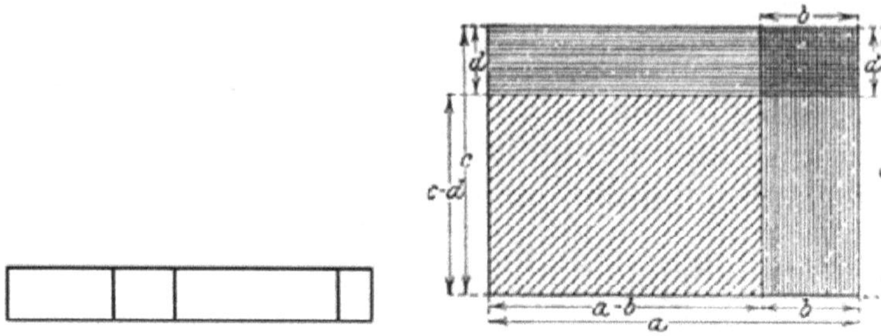

(a) (b)

Fig. Visual proofs of basic rules of arithmetic

As an additional example, the reader may wish to consider how a 'loaf' of balls packed in the shape of a rectangular prism of edges a, b and c may be seen as proving the associativity of multiplication: $a(bc) = (ab)c$. (Hint: One can imagine slices being cut vertically on the one hand and horizontally on the other.)

Tasks Supporting Positional Numeration

I end this brief survey of additional artefacts and tasks by considering a few tasks intended to foster the understanding of positional-value numeration systems (in particular ones in base ten).

Trading Bases

A natural question, when considering a numeration system in a base other than ten, is to consider how to transfer a given (base ten) numeral to the new base, and vice versa. Eventually such a question could be addressed considering arbitrary bases a and b, going directly from one system to the other without transiting via base ten. Necessary artefacts would then be the Pythagorean tables for the bases at stake. Experience shows that if student teachers are left on their own facing such a task, the following three methods will eventually appear:

- Dividing the given number by the grouping order of the target base (computations then take place in the source base).
- 'Exhausting' the number by multiples of powers of the target base (computations also take place in the source base).
- Evaluating the number in the target base (computations now take place in the target base).

Understanding the principles behind each of these methods sheds light on important aspects of numeration.

Paginating a Book

Here are problem-solving tasks providing nice insights into numeration:

- How many digits (i.e. printed characters) will one use in order to paginate a book of, say, 789 pages?
- Reciprocally, if so many characters have been used to paginate a book, how many pages does it have?
- In a similar spirit, how many times does one use the digit 7 when writing down all the numbers from 1 to 99,999?

Factorials and Fractions

Two problems emphasising the role played by the prime factors of the base are:

- By how many 0s does 77! end?
- Moving to rational numbers, one may ask when does a fraction *a*/*b* (given in its lowest terms) correspond to a terminating decimal expansion?

Casting Out Nines

A nice artefact with a long history is the 'casting out nines' procedure for testing the validity of a given reckoning, for instance, the computation of a product. Examining the functioning of this algorithm on the basis of clock arithmetic provides a rich context for entering into the functioning of our numeration system. A nice aspect that can be raised about this test is the issue of 'false positives'. And what about 'casting out threes' or 'casting out elevens'?

More generally, understanding divisibility criteria is a task fostering the understanding of positional numeration.

Conclusion

A central aim of the arithmetic course we propose to prospective primary school teachers is to help them develop a solid 'conceptual understanding' allowing to perceive mathematics not as a mere bunch of facts to be memorised, but rather as a coordinated system of ideas. We hope this way to contribute to the growth of their autonomy and critical analysis skills.

This paper has concentrated on the competency of teachers in mathematics (and especially in basic arithmetic), a crucial aspect of their preparation – but, it goes without saying, not the whole story (see Section). The interested reader will find in Hodgson and Lajoie (2015) brief comments about how the approach to whole number arithmetic described here can serve as a basis for other numerical contexts (***Z*** and ***Q***), as well as a discussion on how the didactical component of the preparation of primary school teachers can make use of artefacts such as sequences of tallies.

A crucial point raised at different places in Chap. is the fact that artefacts and tasks are intimately linked together. While 'artefacts have the potential to foster students' construction of mathematical concepts in whole number arithmetic' (beginning of Sect.), they do not exist by themselves, pedagogically speaking. They must be related to some mathematical tasks. And reciprocally, as shown repeatedly in this paper, a given mathematical task is typically based on a certain artefact, be it a physical tool, an algorithm, or a device such as a sequence of tallies, coming both with a facet of concrete implementation and one of abstract conceptual object.

In spite of a comment made in the introduction, namely, that the learning of arithmetic by actual pupils is not an immediate aim of the work we do with our student teachers, I would maintain that many of the artefacts and tasks discussed in our arithmetic course (and in this chapter) can be transferred to pupils, but of course with a necessary adaptation, as our target audience comprises adults with already a substantial, even if at times frail, mathematical background and not young children new to such notions.

Acknowledgements I wish to thank Caroline Lajoie and Frédéric Gourdeau for their most inspiring discussions about the topic of this paper. I also wish to express my gratitude to Linda Lessard, sessional lecturer at Université Laval, with whom I have closely collaborated over a period of more than 35 years in the teaching of arithmetic and geometry to prospective primary school teachers.

This paper is dedicated to the memory of William S. Hatcher (1935–2005), my mentor in the field of mathematical logic and former colleague, who influenced in a significant way the basic vision of arithmetic developed in our courses supporting primary education.

References

Ball, D. L., & Bass, H. (2003). Toward a practice-based theory of mathematical knowledge for teaching. In E. Simmt & B. Davis (Eds.), *Proceedings of the 2002 annual meeting of the Canadian Mathematics Education Study Group/Groupe canadien d'étude en didactique des mathématiques* (pp. 3–14). Edmonton AB: CMESG/GCEDM.

Bednarz, N. (2012). Formation mathématique des enseignants: état des lieux, questions et perspectives. In J. Proulx, C. Corriveau, & H. Squalli (Eds.), *Formation mathématique pour l'enseignement des mathématiques: pratiques, orientations et recherches [Mathematical preparation for the teaching of mathematics: Practices, orientations and researches]* (pp. 13–54). Québec: Presses de l'Université du Québec.

Buisson, F. (1911). *Nouveau dictionnaire de pédagogie et d'instruction primaire.* Entrée: Abstraction [*New dictionary of pedagogy and primary education.* Headword: Abstraction]. Paris: Hachette. Retrieved from www.inrp.fr/edition-electronique/lodel/dictionnaire-ferdinand-buisson/document.php?id=1971

Cassidy, C., & Hodgson, B. R. (1982). Because a door has to be open or closed… *Mathematics Teacher*, *75*(2), 155–158. (Reprinted in S.I. Brown & M.I. Walter (Eds.), *Problem posing: Reflections and applications* (pp. 222–228). Hillsdale: Lawrence Erlbaum, 1993.)

Courant, R., & Robbins, H. (1947). *What is mathematics? An elementary approach to ideas and methods.* New York: Oxford University Press.

Davis, M. (1958). *Computability and unsolvability*. New York: McGraw-Hill Book Company.

Euler, L. (1738). *Einleitung zur Rechen-Kunst zum Gebrauch des Gymnasii bey der Kayserlichen Academie der Wissenschafften in St. Petersburg*. [*The Art of Reckoning*.] St. Petersburg, Russia: Academischen Buchdruckerey. Reprinted in Euler's *Opera Omnia*, ser. III. vol. 2 (pp. 1–303). Leipzig, Germany: B.G. Teubner, 1942. Retrieved from www.math.uni-bielefeld.de/~sieben/euler/rechenkunst.html

Freudenthal, H. (1973). *Mathematics as an educational task*. Dordrecht: D. Reidel Publishing Company.

Grossman, P. L., Wilson, S. M., & Shulman, L. S. (1989). Teachers of substance: Subject matter knowledge for teaching. In M. C. Reynolds (Ed.), *Knowledge base for the beginning teacher* (pp. 23–36). Toronto: Pergamon Press.

Hodgson, B. R. (2004). The mathematical education of school teachers: A baker's dozen of fertile problems. In J. P. Wang & B. Y. Xu (Eds.), *Trends and challenges in mathematics education* (pp. 315–341). Shanghai: East China Normal University Press.

Ifrah, G. (2000). *The universal history of numbers: From prehistory to the invention of the computer*. New York: Wiley.

Kleene, S. C. (1952). *Introduction to metamathematics*. Amsterdam: North-Holland Publishing Co.

Klein, F. (1932). *Elementary mathematics from an advanced standpoint: Arithmetic, algebra, analysis*. New York: Macmillan. [Translation of volume 1 of the three-volume third edition of *Elementarmathematik vom höheren Standpunkte aus*. Berlin, Germany: J. Springer, 1924–1928.]

Lemay, F. (1975). L'expression numérique du plural (méthode des orbites). [The numerical expression of the plural (the method of orbits).] Laboratoire de didactique, Faculté des sciences de l'éducation (Monographie no. 8). Québec: Université Laval.

Lorenzen, P. (1955). *Einführung in die operative Logik und Mathematik*. Berlin: Springer.

Ma, L. (1999). *Knowing and teaching elementary mathematics: Teachers' understanding of fundamental mathematics in China and the United States*. Mahwah: Lawrence Erlbaum Associates.

OED (2016). *Oxford English Dictionary* (online version). Headword: Score—II ('Notch cut for record, tally, reckoning'), 9.a.

Steen, S. W. P. (1972). *Mathematical logic with special reference to the natural numbers*. Cambridge: Cambridge University Press.

Swetz, F. J., & Katz, V. J. (2011). *Mathematical treasures*—Margarita philosophica *of Gregor Reisch*. Convergence. Retrieved from www.maa.org/press/periodicals/convergence/mathematical-treasures-margarita-philosophica-of-gregor-reisch

UNESCO. (1973). *New trends in mathematics teaching* (Vol. III). Paris: UNESCO.

von Neumann, J. (1923). Zur Einführung der transfiniten Zahlen. *Acta scientiarum mathematicarum, 1*, 199–208. English translation in J. van Heijenoort (Ed.) *From Frege to Gödel: A source book in mathematical logic, 1879–1931* (pp. 346–354). Cambridge, MA: Harvard University Press.

Wittenberg, A., Soeur Sainte-Jeanne-de-France & Lemay, F. (1963). *Redécouvrir les mathématiques: exemples d'enseignement génétique*. [*Rediscovering mathematics: Examples of genetic teaching*.] Neuchâtel: Éditions Delachaux & Niestlé.

Wittmann, E. (1975). Natural numbers and groupings. *Educational Studies in Mathematics, 6*(1), 53–75.

8

Some International Perspectives: How to Teach and Assess Whole Number Arithmetic

Jarmila Novotná, Berinderjeet Kaur
Ann Gervasoni, Mike Askew, Michiel Veldhuis, Catherine Pearn,
and Xu Hua Sun

Introduction

About the Chapter

In accordance with the programme components at the ICMI Study 23 Conference in Macao (Theme 4), this chapter focuses on the diverse theoretical and methodological frameworks that capture the complex relationship between whole number learning, teaching and assessment. Its aim is to bring these diverse perspectives into conversation.

The importance of the theme for students' development of understanding mathematics, and their overall learning, is obvious. This is highlighted in the Discussion

---

Electronic Supplementary Material The online version of this chapter (doi:10.1007/978-3-319-63555-2_11) contains supplementary material.

The first two authors wrote the chapter with the collaboration of other working group members.

The first author was supported by the project PRVOUK P15 *The school and teaching profession in the context of growing educational demands*.

The last author was supported by Research Committee, University of Macau, Macao, China (MYRG2015-00203-FED). The opinions expressed are those of the author.

J. Novotná (✉)
Charles University, Prague, Czech Republic, and Laboratoire "Culture et Diffusion des Savoirs", Université de Bordeaux, Bordeaux, France
e-mail: jarmila.novotna@pedf.cuni.cz

B. Kaur (✉)
National Institute of Education, Nanyang Technological University, Singapore, Singapore
e-mail: berinderjeet.kaur@nie.edu.sg

A. Gervasoni
Monash University, Melbourne, Victoria, Australia

Document (this volume, Appendix 2) as well as in other chapters of the book. There are many approaches to WNA teaching and assessment throughout the world and within countries. Describing all approaches cannot be presented in one chapter of the book. Instead, the authors present one concrete example of school practice – the lesson observed by ICMI23 participants at a primary school in Macao – and use the lesson as a stimulus for discussing ideas about teaching, learning and assessing WNA.

The chapter is divided into seven sections (including this introductory text with the overview of the Theme 4 programme). The focus for all sections is how teachers promote the development of students' metacognitive strategies during their learning of WNA. Each section develops an important aspect of the theme. A description of the Macao Primary School lesson is first presented (Sect.). This provides the context for the subsequent sections, except for the one focusing on textbooks (as during the lesson textbooks were not explicitly used). The two versions of the variation theory (Sect.) and the theory of didactical situations (Sect.) are used as lenses to interpret the lesson. How teachers' knowledge is related to their teaching approaches is discussed in Sect. , while Sects. focus, respectively, on assessment and textbooks which are significant domains for the teaching and learning of WNA.

The topics discussed in this chapter each have important implications for teacher education. These are discussed in Chapter. Furthermore, there were several ques tions from the Discussion Document (this volume, Appendix 2) discussed during the working group 4 sessions and published in the conference proceedings. In order to show the richness of discussions among the working group 4 participants, the descriptions of all contributions to the topic are summarised in Section. Readers interested in more detailed information about any discussed aspect can find the cor-responding papers in the ICMI Study 23 Proceedings.

What Was Presented at the Conference: Overview

Theme 4 addressed general and specific approaches to teaching and assessing WNA. In the thematic group, theoretical and methodological frameworks that could capture the complex relationship between whole number learning, teaching and

---

M. Askew
University of the Witwatersrand, Johannesburg, South Africa, and Monash University, Melbourne, Victoria, Australia

M. Veldhuis
Utrecht University, Utrecht, The Netherlands

C. Pearn
The University of Melbourne, Melbourne, Victoria, Australia

X.H. Sun
Faculty of Education, University of Macau, Macao, China

assessment were considered. The background question for all contributions to Theme 4 was the same as for this chapter: how can teachers promote the development of students' metacognitive strategies during the learning of WNA?

Fourteen papers written by authors from 13 countries were accepted for this theme. They addressed the issues of teaching and assessing WNA from different perspectives. The contributions by participants from many countries offered a unique opportunity to compare and contrast different approaches to teaching and assessing WNA.

For presentation and discussion, the papers accepted for Theme 4 were divided into five subgroups according to their main focus. The summary that follows uses these subgroups as an organisational structure. We are aware that it is not possible to make disjoint groups of papers when attending to their main focus. Note: there is no ranking in the order of topics.

Teaching Approaches

Askew (2015) focused on place value in Grade 2 in South Africa. He argued that within a context focused around whole-class teaching, it is still possible to engage learners with mathematics in ways that go beyond merely reproducing procedures demonstrated by the teacher.

Cao et al. (2015) presented the characteristics of the Chinese traditional approach from the perspectives of content, organisation, the arrangement of teaching, ways of presenting and cognitive demand level with a special emphasis given to multiplication tables.

Knowledge of Teachers

Ekdahl and Runesson (2015) examined shifts in the nature of responses of three South African Grade 3 teachers to students' incorrect answers when teaching the part-whole relationship in additive missing number problems and discuss consequences.

Lin (2015) focused on teaching the structure of standard algorithms for multiplication with multi-digit multipliers via conjecturing as one of effective instructional approaches to teaching multiplication with multi-digit multipliers.

Barry et al. (2015) investigated the variables that determine the difficulty of an additive problem. They showed that the knowledge of variables that determine it differs considerably from one teacher to another. They justified that the roots of these differences are in the differences of teachers' pedagogical beliefs.

Curriculum

This group of contributions is not fully homogeneous. It includes papers focusing on WNA curricula in different countries, on the outcomes when applying curricula and on the influence of using different types of textbooks.

Kaur (2015) presented the primary school mathematics curriculum in Singapore, focusing on the model method, a tool for representing and visualising relationships. In Singapore, it is a key heuristic that students use for solving WNA word problems.

Wong et al. (2015) reported on Macao's 15 years of experiences of primary mathematics education, after the official handover of the former Portuguese enclave to China in 1999. They argued that no educational system can provide Macao with a ready-made curriculum model.

Sensevy et al. (2015) analysed the principles and rationale of a curriculum for WNA teaching in the first grade in France.

Brombacher (2015) reported on the situation in Jordan. He suggested that deliberate and structured daily focus on foundational whole number skills can support the development of children's ability to do mathematics with understanding.

Textbooks

Alafaleq et al. (2015) examined how equality and inequality of whole numbers are introduced in primary mathematics textbooks in China, Indonesia and Saudi Arabia.

Zhang et al. (2015) investigated four sets of primary mathematics textbooks used in Hong Kong using content analysis.

Assessment and Evaluation of WNA

Zhao et al. (2015) discussed challenges that Chinese primary teachers faced when supposed to implement classroom assessment techniques.

Gervasoni and Parish (2015) presented the results of one-to-one assessment with nearly 2000 Australian primary school students. They highlighted the challenge of meeting each student's learning needs, and demonstrated the complexity of classroom teaching.

Pearn (2015) compared the reactions of the Grade 4 teachers in one school (with the same curriculum) with their students' results on a WNA test.

The Discussion in the Working Group

The eight 1-hour sessions in the working group were organised in two different forms. The first five sessions were devoted to the discussion of the participants' accepted papers. The papers were grouped according to their topics dealt with: teaching approaches, knowledge of teachers, curriculum, curriculum and textbooks and assessment of WNA. Then three discussion sessions followed. Their themes of discussion were: teachers' knowledge of their students' understanding, performance and personality traits that can help teachers plan for effective teaching strategies;

multicultural approaches and traditions of teaching WNA; the role of textbooks and artefacts in teaching WNA; and the role of custom-made curriculum to improve the learning of WNA. The work was organised partly in the mixture of small group and plenary discussions.

The programme was enriched by the contribution of Marianela Zumbado Castro from Costa Rica (CNAP). Her presentation covered the information about the important development of number meaning in the Costa Rican mathematical programme. She highlighted approaches to calculations and approximations and the use of multiple representations in troubleshooting. These domains were covered from an eminently pragmatic perspective that emphasised student action. She emphasised that numbers occupy a big place from first to sixth graders, in order to promote mathematical processes and positive attitudes for its proximity to the context and its close connection with the other areas.

It is obvious that the topics dealt within WG4 were not disjoint with other WGs. There were overlaps with other working group topics as well as with the panel about teacher education. However, the common topics were tackled from different perspectives in each part of the programme and can be seen as complementing each other.

A Mathematics Lesson Focused on Addition Calculations with Two-Digit Numbers at a Primary School in Macao

Introduction

Teachers in Macao in common with primary school teachers across the world are working to transform their approach to teaching mathematics. They have a strong focus on students learning mathematics with deep conceptual understanding and developing creative thinking. In this chapter, we aim to explore issues in the teaching of arithmetic with reference to an illustrative lesson that we observed at a primary school in Macao in June 2015. The Grade 1 class we observed were all enthusiastic and well focused on the activities and learning throughout the lesson. This school, we were told, aims to ensure that the students will be well behaved, enjoy their lives and be good at creative thinking for the future. The mathematics team of 14 teachers was described by the principal as progressive and innovative. They meet together in different levels at least once per week to share their objectives and activities. They also engage in research practices focused on school-based studies of teaching aimed to promote positive changes in teaching and learning. Every teacher is responsible for teaching one demonstration lesson and observing 20 lessons in an academic year so that the mathematics teachers can learn from each other. The school also organises training for elite mathematics teams from the kindergarten to the grade 9 who join mathematics competitions. This activity aims to enhance logical thinking and project-based practice.

The Lesson

The Grade 1 students we observed have five mathematics lessons per week and one mathematics reading lesson per week (a mathematics reading lesson involves stories and written problem-solving, mathematical games, project-based activities and hands-on activities). The lesson began with a *lead in* stage during which students engaged in 3 min of mental calculation practice. The lesson proceeded with several stages during which the students explored a range of situations and strategies for adding two-digit numbers. Throughout the lesson, the teacher circulated to observe and discuss the students' strategies and to select students to describe and explain their calculation strategy to the whole class. In one stage of the lesson, the students worked in fours to use models to represent the calculation and solution. A variety of real-world calculations and bare number calculations were investigated and solved and discussed. The Grade 1 students we observed were all enthusiastic and well focused on the activities and learning throughout the lesson (Electronic Supplementary Material: Sun 2017b).

The setting and sequence of activities of the lesson are summarised in Table shown below.

Table Setting and sequence of activities of the lesson

Seating organisation for the lesson	See Fig.
The class of 22 Grade 1 children (6-year-olds) sat in pairs at their desks.	
Introduction and welcome	See Fig.
The lesson began with a warm greeting from the teacher and a respectful bow from the students.	
Stage 1 lead in: fluency with number facts	See Fig.
The first activity in the lesson (the lead in stage) focused on the students practising number facts – number combinations for ten. The students were shown a set of number facts to solve and recorded their answers. This practice was timed with students recording when they completed all the calculations. The teacher roved around and observed the students as they worked.	
The answers were corrected as a group with the stated aim being for the students to improve their previous score and time. The teacher enthusiastically encouraged the students to improve and asked them to state if they had improved their time and score.	
Stage 2 situation setting: addition of a two-digit number and a one-digit number without regrouping – using pictorial representations of realistic contexts	See Fig.
In this section of the lesson (situation setting), the teacher presented a realistic situation for the students to explore. The first problem showed a pictorial representation of four packets of ten candies and three single candies. They needed to work out the total: 40 + 3 = 43.	

(continued)

Table (continued)

The next situation presented was a pictorial representation of candies organised onto trays of ten and loose candies. The calculation was 25 + 2 = 27. Once the problem was solved the teacher guided a discussion about the calculation strategies used. The photo shows splitting the 25 into tens and ones and then grouping the ones to calculate the total easily.	See Fig.
Finally, the students considered a situation involving pencils organised into boxes of ten and some loose pencils. The calculation involved two-digit numbers. The solution discussed by the teacher and students considered grouping the boxes of ten together and adding the number of loose pencils.	See Fig.
25 + 20 = 45	
20 + 20 + 5 = 45	
Stage 3 group counting and sharing: addition with two-digit numbers and one-digit number (with regrouping ones)	See Fig.
This stage of the lesson was described as the group counting and sharing stage during which the students were expected to communicate, conceptualise and inquire. The context was 'party time'; materials (candies) were used to represent the situation and the calculation strategies for 24 + 9 = 33. The students worked in groups of four to discuss the possible strategies for this calculation, while the teacher moved between the groups to observe, listen in and discuss the strategies. Students were selected to come to the front of the class and explain their strategies using the document camera and interactive white board. The teacher explored three different strategies with the class. She had anticipated these strategies before the lesson and prepared the posters for display.	
23 + (1 + 9) = 33	
20 + (4 + 9) = 33	
(24 + 6) + 3 = 30 + 3	
Stage 4 Practice Stage: calculation competition	See Fig.
This stage of the lesson had three components. During Practice Stage 1, the student found solutions to a set of problems while the teacher circulated to observe and discuss the strategies. The teacher again selected several students to come to the front to explain their calculation strategies. The teacher highlighted carefully the different strategies used. However, after only completing one of the four calculations on the board, the teacher moved to Stage 2. In Practice Stage 2 the students worked in pairs and were asked to choose their favourite digits from an envelope to make a new two-digit number. The teacher gave them another card with the addition sign and a single digit number (9 followed by 7). The students had to add the one-digit number to their two-digit number. Several strategies were shared with the class.	
One child explained a calculation that bridged 100: 95 + 9 = 104. She had some difficulty describing her strategy to the class and the teacher assisted her.	See Fig.
Practice Stage 3 was a return to the calculating competition (Practice Stage 1) where the students were encouraged to share their answers and their strategies.	See Fig.
Lesson summary	
The conclusion of the lesson was a summary by the teacher about what they had explored and learnt during the lesson. The teacher emphasised that the students could use the 'making 10' strategy to add a two-digit number and a one-digit number.	

Fig. Seating organisation

Fig. Introduction and welcome

Fig Lead in: fluency with number facts

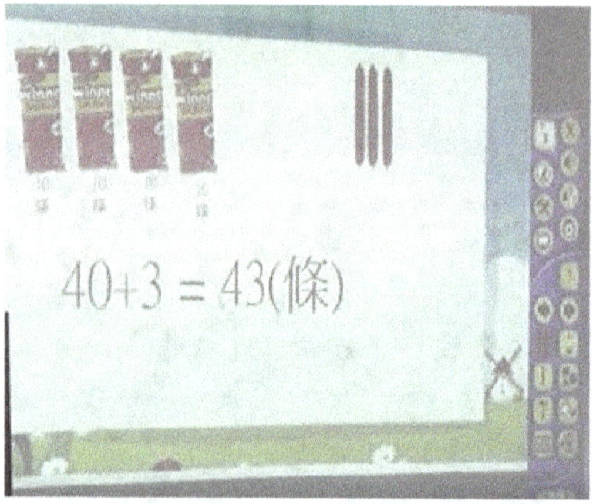

Fig. First situation setting

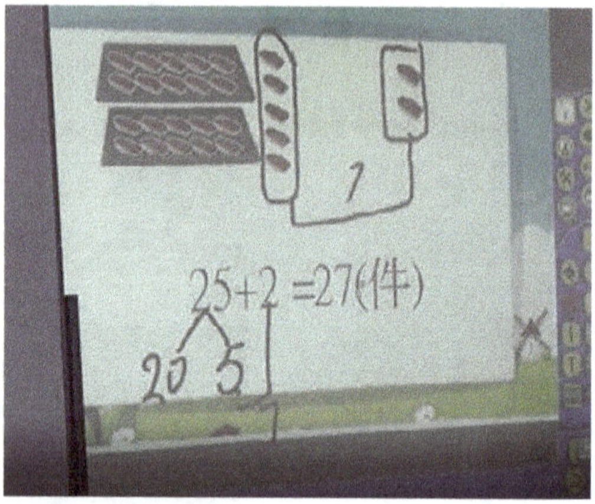

Fig. Second situation setting

Fig. Third situation setting

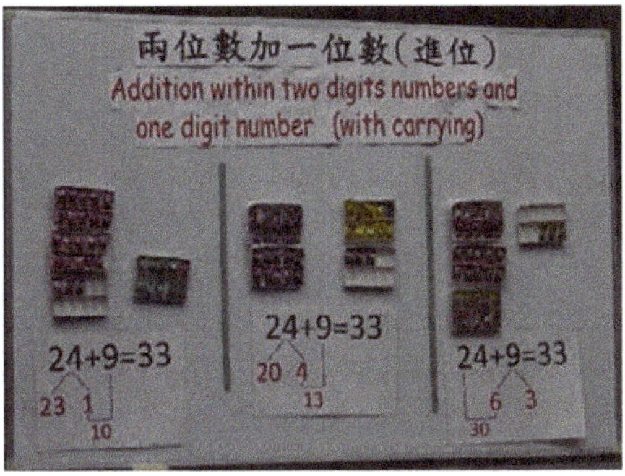

Fig. Group counting and sharing

Fig. Practice stage 1

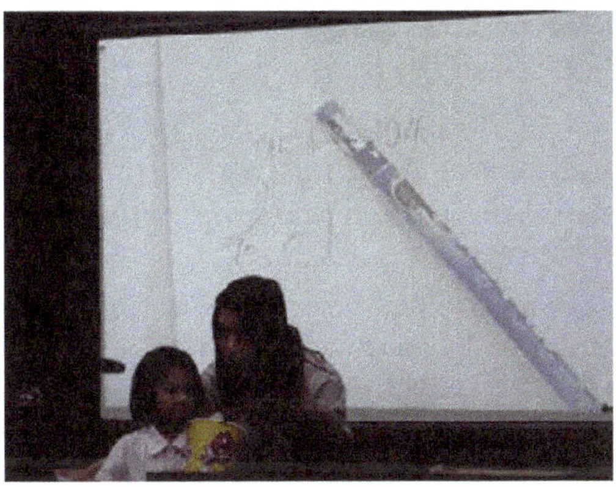

Fig. Practice stage 2

Fig. Practice stage 3

The Impact of Teachers' Knowledge of Pedagogy, Learning Trajectories, Mathematics and Students (Cognitive, Social, Emotional, Context, etc.) on Children's Learning of Whole Number Arithmetic

Introduction

Many everyday calculations involve adding combinations of one-digit and two-digit numbers. This is so whether we are shopping, measuring for sewing or a building project or analysing data in a rainfall graph. This partly explains why adding one- and two-digit numbers is a key focus in primary school mathematics education. However, approaches to teaching this type of calculation vary within and between countries. The teaching approach in many countries today emphasises children learning a variety of arithmetic strategies so that they can successfully and efficiently perform calculations mentally. However, other approaches draw upon text-books that focus on children learning standard written methods for performing these calculations. It is important for teachers to understand the affordances and limitations of various teaching approaches for children's arithmetic learning.

Responding to Children's Current Knowledge of Whole Number Addition and Subtraction

The Macao lesson vignette described earlier in this chapter provides an example of an approach to teaching children to perform calculations with one-digit and two-digit numbers that would be recognised in many countries (e.g. Singapore, Germany, Australia, Canada, Thailand). The teacher used her knowledge (assessment) of the students' current mathematics knowledge and pedagogy to select tasks with an appropriate level of challenge and that aligned with the curriculum expectations.

She made a concerted effort to motivate the students to engage in the lesson activities and work hard to improve their knowledge, and she was encouraging of the students' efforts to learn. The teacher provided the children with time to work on their own, time to work in small groups to discuss ideas and solutions and time to observe and listen to the solutions presented by several students. This suggests that she may be influenced by social-cultural perspectives (e.g. Vygotsky 1980) in designing the lesson structure. However, this may be the teaching norm in her school community. With respect to the tasks, the teacher provided a balance between tasks that involved bare number calculations and those that were connected to everyday situations. There was also the opportunity in the lesson for children to use materials and pictures to model their solutions processes to support their mathematical reasoning. There was an overarching focus on the children understanding the solution strategies they were using or viewing. However, in contrast to the approach in Singapore (Kaur 2015) and advocated by Bruner (1960), there was little use of materials being used to assist individual children's learning progress from arithmetic reasoning based on concrete representations through to pictorial and abstract representations. In contrast, the approach in the Macao lesson required all the children to engage with concrete, pictorial and abstract models at set points in the lesson. There was no sense that the teacher was using formative assessment to select concrete, pictorial or abstract models for individual children based on their current understanding or to encourage the children to select these different representations themselves.

One set of tasks in this lesson involved the students performing calculations with numbers such as 25 + 9. Pictures of snacks in boxes of ten and loose ones were used to represent the quantities and to model the various solution processes that were predetermined by the teacher. This set of solutions was elicited from the children and discussed as a class group to build understanding and provide children with examples of strategies they may not have spontaneously considered. This enabled the class to discuss the advantages of various strategies in relation to the numbers involved. This teaching approach of inviting children to demonstrate and discuss various solutions is widely used in countries such as Japan (Murata and Fuson 2006), the Netherlands (RME; van den Heuvel-Panhuizen and Drijvers 2014), Germany (Selter 1998) and Australia (Clarke et al. 2002).

One advantage of all the children in a class performing the same calculations is that they can all participate in a meaningful discussion about the various arithmetic strategies used. However, tasks such as 25 + 9 do little to develop creativity, challenge or persistence. An alternative approach is to use an open task that enables children to create and discuss a range of solutions. For example, the students may be asked to create a set of solutions for the task, 'Ivy added two numbers and the sum was more than 32. What might the numbers be?' This open task encourages the students to persist in producing a range of solutions that are creative and complex enough to challenge them to think hard. The solutions and arithmetic strategies can then be discussed as a class to extend the understanding of all. This approach is advocated by Sullivan et al. (2015).

The challenge for a teacher during a lesson such as the one described earlier is to plan how the learning may be differentiated and how the lesson can develop creativity, which was a stated aim of Hou Kong Primary School in Macao and is highlighted as an important goal of mathematics education in many international settings. In the observed lesson, it seemed that all students were able to successfully perform the calculations, but some were clearly more confident than others, and some children may have been able to solve more complex problems, given the opportunity. However, the tasks and teaching actions were not adapted to respond to any formative assessment data collected by the teacher during her observations or discussion with children. This issue of task differentiation and developing creative mathematical thinking are important considerations when considering highly effective teaching of whole number arithmetic. Further, in many countries, where class sizes are very large, the differences in the knowledge of the students may pose the teacher with challenges. Thus differentiating tasks and instruction is an important role for the teacher when children are learning whole number arithmetic.

Differentiating Instruction Based on Children's Current Knowledge

Children's development of heuristic strategies for addition and subtraction has been well described (e.g. Steffe et al. 1988; Murata and Fuson 1997). Such research formed the basis for describing the six addition and subtraction strategies growth points established during the Early Numeracy Research Project (Clarke et al. 2002) to describe children's learning in this domain. Growth points are established for a child following assessment by the teacher using a detailed scripted one-on-one assessment interview (Clarke et al. 2002). This is a common assessment approach in Australia and New Zealand (Bobis et al. 2005). A feature of assessment interviews is that they enable the teacher to observe children as they solve problems to determine the strategies they used and any misconceptions (Gervasoni and Sullivan 2007). They also enable teachers to probe children's mathematical understanding through thoughtful questioning (Wright et al. 2000) and observational listening (Mitchell and Horne 2011).

Growth point results for nearly 2000 Grade 1 to Grade 4 students (Fig.) collected during the Bridging the Numeracy Gap Project (Gervasoni et al. 2011) confirm that there is a wide distribution of children's addition and subtraction strategies growth points in each class. These data indicate that 96% of Australian Grade 1 children, 75% of Grade 2 children, 46% of Grade 3 children and 30% of Grade 4 children used counting-based strategies for calculations, such as 4 + 4 and 10 – 3. The fact that so many Grade 4 children remain reliant on counting strategies and concrete models for calculating and that almost no Grade 4 students could solve mental calculations involving two-digit and three-digit numbers (growth point 6) is at odds with the tasks typically found in Grade 4 textbooks that involve calculations

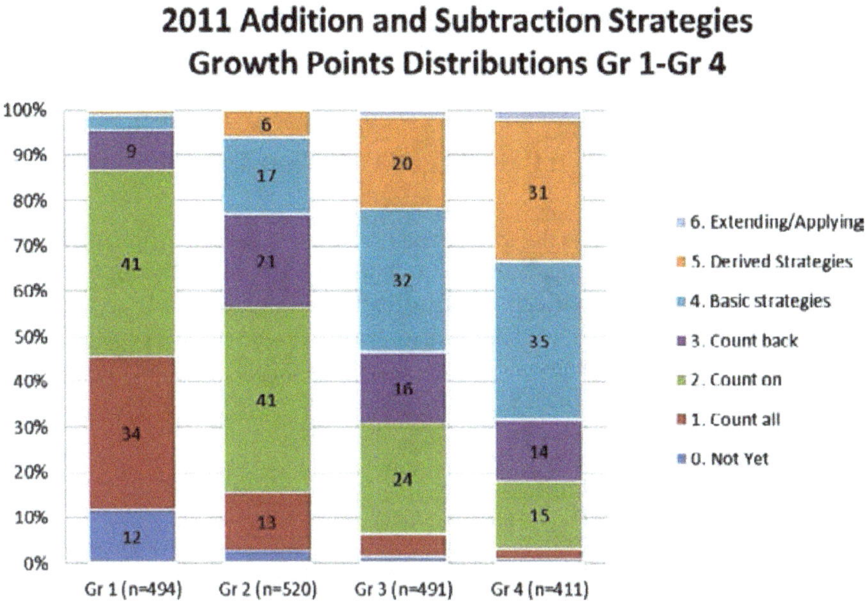

Fig. Addition and subtraction growth point distribution for Grade 1–4 children

with much larger numbers. Data such as these demonstrate the need for teachers to understand the current knowledge of students through observation of their calculation strategies, understand the typical developmental pathway in this domain and understand the teaching strategies that respond to the individual needs of students so that they can differentiate tasks. This is not possible if teachers only use written tests (Clements and Ellerton 1995) or one textbook for a class that does not differentiate learning and extend understanding. Rather, if teachers observe that a child uses a count-all strategy (growth point 1) and refer this to a learning framework, then they can appreciate that a child's shift to using a count-on strategy (growth point 2) or abstract reasoning strategies (growth point 4) requires the teacher to hide some of the physical models to prompt children to produce a mental image that enables them to count-on or reason abstractly.

In summary, there is no single 'formula' for describing children's whole number knowledge or the instructional needs of children in a particular grade. Meeting the diverse learning needs of children requires teachers to be knowledgeable about how to identify each child's current mathematical knowledge and how to customise their teaching accordingly. This calls for rich initial and formative assessment tools capable of revealing the extent of children's whole number knowledge and calculation strategies and an associated framework of growth points capable of guiding teachers' curriculum and instructional decision-making. Assisting children to learn mathematics is complex, but teachers who are equipped with the pedagogical knowledge and actions necessary for responding to the diverse needs of individuals are able to provide children with the opportunities and experiences that will enable them to thrive mathematically.

View of Lesson through the Lens of Variation Theory

Variation theory has come to the foreground in the last decade in different parts of the world (Huang et al. 2006) as a lens through which the classroom learning can be designed, described and analysed. First we look at the 'indigenous' approach to variation problems (Sun 2011a, b, 2016), in order to give an insider's perspective on the lesson (design and functioning), and second we situate the lesson within the international debate (e.g. Marton et al. 2004) about the observed lesson.

The Insider's Perspective: 'Indigenous' Variation Practice

In this section, we examine the lesson through a lens of 'indigenous' variation practice as a means of providing the insider perspective, in order to enhance comprehensive understanding of the lesson. In this section, we mainly focus on the situation of the candy box problem. In order to understand the classroom process, it is necessary first to describe the past experience of the students, which followed the Chinese curriculum as it is described in Chinese textbooks. The importance of textbooks in Chinese curriculum must be emphasised, as teachers follow the textbooks' directions (which draw on the standards) very carefully. As said in the previous sections, this lesson took place close to the end of the school year; hence, students were already familiar with the previous content.

The Prior Students' Knowledge

The students' previous knowledge about different cases of addition and subtraction had been developed, fostering grouping, regrouping and ungrouping, following the Chinese tradition. Consider the following examples of activities taken from their textbook for the first months of the first grade (Fig.).

The image on the left provides a good example of *explicit variation*. In the first line, there is the standard notation for decomposing 10 in different ways. Then, in each of the following lines, there is a scheme (the problem situation, given in iconic form) referring to one of the above decomposition. Each problem is interpreted in different ways, according to the indigenous variation practice (OPMC, i.e. one problem, multiple changes, Sun 2011b), hinting at addition and subtraction together. The same problem is focused at the same time with different interpretations, creating a

Fig. Examples introducing the decomposition of 10 in the Chinese textbook (Mathematics Textbook Developer Group for Elementary School, 2005, vol. 1)

very strict link between addition and subtraction and a focus on algebraic thinking (relations between numbers). In the same image, another explicit variation appears increasing the quantity of missing numbers to be detected and changing their position in the mathematical expressions. This is expected to be done in the same lesson as one textbook page is used in one lesson. This approach is very different from other ones practised in some Western countries, where addition and subtraction may be taught separately (see Sun et al. 2013 for Portugal; Bartolini Bussi et al. 2013 for Italy).

The image on the right exploits the standard notation for decomposing 10 in order to solve, in different ways, the addition 9 + 5 with regrouping (OPMS, i.e. one problem, multiple solutions, Sun 2011b).

Besides that, students know already how to add tens and how to add two-digit numbers without regrouping. Hence, the students knew what was necessary to handle the situation of the day, concerning the candy boxes and the addition with regrouping 24 + 9.

The Lesson Plan: The Situation of the Day

According to the framework of Chinese open classes, the lesson plan was given to the international observers a few days in advance. It contained details about the organisation, the teaching topic (addition within two-digit numbers and one-digit number with regrouping), the learning objectives and the students' previ-ous knowledge (Bartolini Bussi and Sun 2015).

The situation of the day was the problem of the candy boxes: it was carefully described in the lesson plan (excerpt) (Table).

Table The lesson plan (excerpt)

Situation setting	(a) Teacher gives a situation to the class. 'There are many guests in our school today. So, Miss Amanda prepares some food for them. Class, can you help me count the food as fast as you can?'	15 min
Problem solving	(a) Students work in groups.	
	(b) Provide some candies to each group and let them count.	
Group counting and sharing *(communication, conceptualising, inquiring)*	(a) T invites some group to report their finding about how to count the candies altogether.	
	(b) Give comments to the groups and use the multimedia to show three different ways to count the candies. *The first way of putting candies altogether*	
	There are 24 candies on the left, and then there are 9 candies on the right.	
	Next, encourage students to investigate and move 4 candies on the left and 6 candies on the right to 'making 10'. Finally, 30 candies plus 3 candies equals 33 candies altogether.	
	The second way of putting candies altogether	
	There are 24 candies on the left, and then there are 9 candies on the right.	
	Next, encourage students to investigate and move 1 candy on the left and 9 candies on the right to 'making 10'. Finally, 23 candies plus 10 candies equals 33 candies altogether.	
	The third way of putting candies altogether	
	There are 24 candies on the left, and then there are 9 candies on the right.	
	Next, encourage students to investigate and move 4 candies on the left plus 9 candies on the right equals 13 candies. Then, there is a 'making 10' in 13. Finally, 20 candies plus 13 candies equals 33 candies altogether.	

The three mentioned ways of 'putting candies altogether' hint at different ways of 'making 10':

- To increase 24 to 30 (3 tens).
- To increase 9 to 10 (1 ten).
- To decrease 24 to 20 (2 tens).

The teacher knew in advance that some students might have difficulties; hence, she was ready to encourage students, forcing the grouping strategy, using the concept of 'making 10'.

The Functioning in the Classroom

The time planned for the solution was very short (15 min). However the students succeeded in solving the problem finding the three solutions anticipated by the teacher in a shorter time. The process may be outlined looking at the transcript (translated from Cantonese). The final solutions produced by the groups are in the figure of the Stage 3.

00.00 Here you have some candies, we can count how many candies, each group has candies boxes. Put on the table and calculate. Think how to use the candies and find the result. You can split candies and move between the boxes on the left and on the right.

... Students work in small groups.

02.14 Have you finished? Good job! And you? Have you finished? And you? Good job!

02.32 Very fast! Very good!

02.39 Let's look together. Turn to this direction (the whiteboard).

... The first solution: $24 + 9 = 23 + (1 + 9) = 33$

03.07 Let's look how she did. Hang to the board. Tell them how and why you did this way. To take away the group of 4 and put there. Why? I understand, in this way you have on the right the sum 10, on the right is three groups of 10, 30, right? All together is 33.

04.00 Do you have different methods? Did you think another way? Raise your hands. Did you think another way? Please, come here, carry your boxes.

... The second solution is given: $24 + 9 = 20 + (4 + 9) = 33$

06.05 Good job! Are there other methods? Have you thought in a different way?

... The third solution: $24 + 9 = (24 + 6) + 3 = 33$

08.03 Many solutions, many different solutions. Are there different solutions? How did you move the candies? This is the same as the first method. Other solutions? How did you do? It's the same. You also. Other solutions? Do you have other methods? Let me know

09.00 In mathematics is there only one solution?

Voices No no no.

We may have many different solutions. Yes, you have made different solutions for this problem. To make the addition of a two-digit number and a one-digit number we may use different ways to group and move and get the same result.

Later ...

31.43 In mathematics is there only one solution? You may use any method, you have only to find the same result and you look for the fastest method.

(Electronic Supplementary Material: Sun 2017a, b)

It is worthwhile to observe how the teacher highlights the variation (one problem, multiple solutions (OPMS)) that seems to be the most important global message of the lesson, besides the technique of grouping-ungrouping. Variation of the 'making 10' solution is emphasised and reinforced in other episodes of the lesson, when other problems to be solved orally or in writing are given and are reconsidered at the end (min. 31.43). The coherence between the lesson plan and the realised interaction is very high. According to some scholars (Wang et al. 2015), 'recent international studies in mathematics education have identified a high degree of instructional coherence as a distinguishing feature in classrooms in China' (p. 112). The teacher's discourse showed a high degree of coherence throughout the lesson and, apparently, the students developed a sense of what was going on, showing choral answers to some questions.

A Short Summary

In this section we have tried to give an insider perspective of an 'indigenous' pedagogical practice. The practice, namely, variation problems, is known widely in Chinese mathematics curricula as 'one problem, multiple solutions' (OPMS, 一題多解, varying solutions), 'one problem, multiple changes' (OPMC,一題多變, varying conditions and conclusions) and 'multiple problems, one solution' (MPOS, 多題一解, varying presentations). We have seen examples of the first two above. A more comprehensive summary of this kind of variation problems is in Chapter . A complete discussion and comparison with Marton's variation theory is in Sun (2011b). Variation practice aims at abstracting and generalising, focusing on relationships between numbers rather than on arithmetic operations. This is consistent with the aim of developing algebraic thinking (Cai 2004; Sun 2016). This 'indigenous' pedagogical practice has clear boundaries that allow one to uncover how variation practice has been and is now used in Chinese classrooms and a true operational effectiveness that might be used to transpose it to other contexts (Bartolini Bussi et al. 2013). An example of transposition from China to Italy is discussed in this volume (Chap.).

A Westerner's Perspective: Marton's Variation Theory

It must be emphasised that the following account is a particular 'reading' of the lesson and in particular a reading of deliberate intent behind the selection and structuring of the examples offered to the students to consider. Such a *post hoc* interpretation is not meant to be 'reading the mind' of the teacher – it could well be that the actual rationale behind the choice of examples bears no resemblance to the interpretation presented here. The intent, however, is not to definitively pronounce what was behind the design of the lesson, but to use the lesson as a starting point for thinking about lesson design and how variation theory (in the sense developed by Ference Marton) might be helpful in designing lessons that provide sound opportunities for learning whole number arithmetic and inform the conversation on what mathematical knowledge for teaching needs to be brought to lesson design.

The starting point is the definition of an exercise as put forward by Watson and Mason (2006a), that is, an exercise is 'a collection of procedural questions or tasks'. Thus, rather than examining the separate questions that learners work on in a lesson, these are collectively regarded as contributing towards a set, an exercise, that may be more or less mathematically and pedagogically structured. Thus, the lesson observed had as a major exercise a number of discrete questions all based around adding a single digit to a two-digit number and that the selection, ordering and ways of working within this exercise led to a coherence in the lesson that amounted to more than the sum of the parts.

To examine this, two particular episodes in the lesson are looked at in detail: the working on the problem of 24 sweets plus 9 sweets and the sequence of calculations where learners chose their own two-digit number to work with. Central to this analysis is an exploration both of how the teacher worked with highly interrelated example sequences (Watson and Mason 2006a) and of the ways of simultaneously working with horizontal relationships within examples and vertical patterns between examples (Watson and Mason 2006b).

Variation Theory

Variation theory (VT) as developed by Ference Marton and colleagues (Marton et al. 2004) is a *theory of learning*, not an all-encompassing theory of pedagogy. VT, for example, sheds no light on whether group work is better than individual work or whether physical materials are more useful than pictures or images, although VT theorists acknowledge that these other features of the learning environment are important. The overall aim of VT as developed by Marton and colleagues is to attend to aspects of learning that focus on the specific mathematical content, distinguishing VT from other theories of learning that provide accounts of how learning occurs that are independent of what actual content is expected to be learnt. As a group of teachers and researchers in Hong Kong engaged in 'learning studies' that draw on VT to look at how diverse classroom communities can learn particular content note:

> Contrary to the belief of some educational theorists, therefore, we believe that one simply cannot develop thinking in isolation from the objects of thought. Learning is always the learning of something, and we cannot talk about learning without paying attention to what is being learnt. (Lo et al. 2005, p. 14)

It is beyond the scope of this chapter to go into all the details of VT, but central to the argument here are the constructs of objects of learning, discernment and variation.

VT acknowledges the intentionality of teaching: teaching is always directed towards specific learning ends or, in the language of VT, objects of learning. An object of teaching is never unitary: any teaching activity is always, and inevitably, directed towards at least two objects of learning. There is the direct object of learning – in this lesson, the addition of two numbers. It is this that learners are usually most focused upon as revealed through their answers to the question, 'What did you learn today'? Answer: 'How to add'. But every teaching activity also encompasses one or more general capabilities that are broader than the specific object of learning, for example, interpreting or generalising. These comprise the indirect object of learning (Marton et al. 2004).

According to Bowden and Marton (1998), our learning is a result of what we are able to discern, distinguish. But we can only distinguish between things when there is variation in our experiences:

> When some aspect of a phenomenon or an event varies while another aspect or other aspects remain invariant, the varying aspect will be discerned. In order for this to happen, variation must be experienced by someone as variation. (Bowden and Marton 1998, p. 35)

Variation is the key to being able to discern. As Watson and Mason (2006a) argue, any aspect of a task or situation that it is hoped learners will discern (the lived object of learning) 'is more likely to be discerned if its variation is foregrounded against relative invariance of other features' (p. 98). In other words, what is not varied is as important to attend to as what is varied. So the teacher is in control of the 'enacted object of learning' through the structuring of the 'example space' (Watson and Mason 2005), that is, a collection of examples that fulfil a specific function (Zazkis and Leikin 2007). And equally, if too much is varied, then either nothing may be discerned or inappropriate or irrelevant features attended to. As Runesson (2005, p. 72) notes, 'studies have shown that exposure to variation is critical for the possibility to learn, and that what is learnt reflects the pattern of variation that was present in the learning situation'.

Variation Within the Lesson

The first example in the exercise of adding a single digit to a two-digit number with carrying in the main teaching part of the lesson was based around the party time problem of adding 9 sweets to 24 sweets. Groups of learners were given 24 candies as two full boxes of ten and a box with only four, together with a box of nine. As the groups worked on solving the problem practically, the teacher roamed around noting the different methods and selecting which groups she was going to invite to share their approach with the class. As noted in the account, the teacher had already prepared posters for three strategies and so was, presumably, looking for groups whose strategy fitted with what she had anticipated.

Each group asked to demonstrate their solution started by putting up on the board the two boxes of ten and the four units and the box of nine to the right of these and modelling with the candies what they had done. In this way, three different partitionings of one of the numbers was elicited: partitioning the 24 into 23 + 1 and effectively doing $24 + 9 = (23 + 1) + 9 = 23 + (1 + 9) = 23 + 10$ (although recorded differently, but consistently across the three solutions, by the teacher). Similarly, there was $24 + 9 = (20 + 4) + 9 = 20 + (4 + 9) = 20 + 13$ and $24 + 9 = 24 + (6 + 3) = (24 + 6) + 3 = 30 + 3$.

Here in one example we see careful variation and non-variation. What was kept constant was the calculation and the partitioning of one of the numbers in a way that anticipated creating a multiple of ten to be added to another number. Thus, while the direct object of learning was to calculate the sum, indirect objects implicitly worked with included the associative rule for addition, equivalence and the simplifying of calculations by creating a multiple of ten. All of this, we suggest, was enacted through carefully working horizontally within one example and carefully determined variation within this horizontal working. So while it might be argued that the teacher, in predetermining the solution methods that she sought, limited the opportunity for attending to other things such as learner creativity, from a VT perspective, the control exercised by the teacher here is central to directing what might be discerned by the learners.

The next set of examples in the exercise that we want to examine is the sequence where learners used digit cards to make their own two-digit number. While the teacher made some play of appearing to randomly select a digit for them to add to their number, it seemed from her actions that the pulling out of 9 was not entirely random. Again, as the learners were working, the teacher moved around and chose three to come and show, now using a visualiser, their working (recreating this, not simply showing the final working). The fact that the third example – 95 + 9 – drew gasps of delight from the class suggests again that the choice and sequencing of the three children invited to share their solutions were carefully considered by the teacher. From the perspective of VT, what was now varied was the choice of two-digit number, but adding 9 (and subsequently 7) kept constant. The choice of 9 again not only, in being close to ten, may have encouraged attention to making a multiple of ten, but also only made a small vertical move from the three methods of adding 24 + 9 still displayed on the board. Thus, learners were encouraged to attend to how they can build on what they have just done, rather than having to deal immediately with both numbers being varied.

So although this lesson can be read as being very tightly controlled by the teacher, that control can be read as located in a carefully thought out exercise, a carefully constructed example space, designed to help learners develop effective methods of calculating.

Comparison

When comparing the two approaches to variation theory presented in Sects., there are many similarities. But there can also be found substantial differences, the roots of which can be briefly summarised as follows: Marton's theory is a theory of learning (Kullberg 2010), while indigenous variation practice is a method of task design for teaching. More information about variation theory can be found in Chapter. A comprehensive book on teaching with variation has been published by Huang and Li (2017).

View Through the Lenses of the Theory of Didactical Situations

One efficient tool for analysing teaching episodes is Brousseau's (1997) theory of didactical situations (TDS). In TDS, two different types of knowledge are distinguished: *connaissance* (*a knowing*) and *savoir* (*knowledge*). In the text of this section, we use the English translations of the French terms introduced in Brousseau (1997, p. 72, Editors' note): 'The former refers to individual intellectual cognitive constructs, more often than not unconscious; the latter refers to socially shared and recognised cognitive constructs, which must be made explicit'.

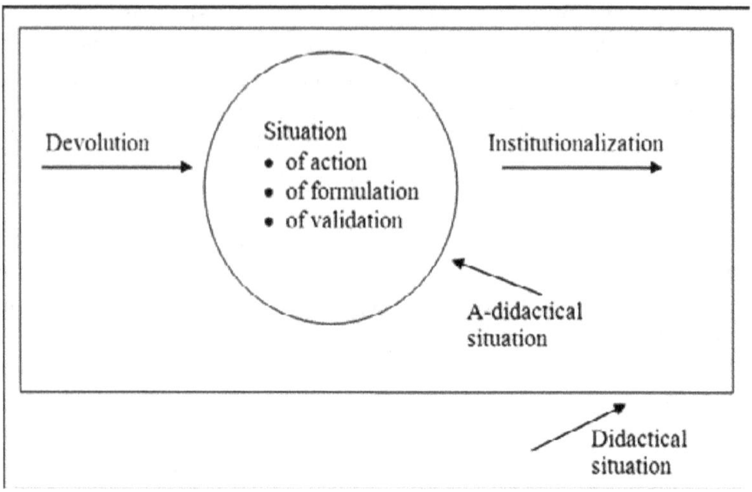

Fig. Scheme of a didactical situation

Didactical Situations

Let us summarise the main characteristics of a didactical situation as used in TDS (Brousseau 1997; Brousseau and Sarrazy 2002). In didactical situations, the teacher organises a plan of action which makes clear her intention of modifying or causing the creation of some knowledge for the student, for example, and which permits her to express herself in actions. Certain didactical situations, the so-called *a-didactical situations*, are (intentionally) partially liberated from the teacher's direct interventions. An a-didactical situation is the autonomous part of an individual or collective activity of students; and student's adaptation to this situation shows the *knowing* involved.

The theory classifies situations according to their structure (action, formulation, validation, etc.) which determines different types of knowledge (implicit models, languages, theorems, etc.). Figure schematically records the structure of a didactical situation.

The a-didactical situation of action is a situation in which a knowing is manifested only by decisions and by regular and effective actions on the *milieu*[1] and where it is of no importance to the evolution of the interactions with the *milieu* whether the actor can or cannot identify, make explicit or explain the necessary knowing.

The a-didactical situation of formulation puts at least two actors into relationship with the *milieu.* Their common success requires that one of them formulates the knowing in question for the use of the other, who needs it in order to convert it to an effective decision about the *milieu.*

The a-didactical situation of validation is a situation whose solution requires that the actors establish together the validity of the characteristic knowledge of this situation.

[1] The word '*milieu'* denotes everything which acts on the student and/or on which she acts.

The institutionalisation of a knowing reveals itself by the passage of this knowing from its role as a means of resolving a situation of action, formulation or validation to a new role, that of reference for future personal or collective uses and, thus, as a piece of knowledge.

Devolution is the process by which the teacher manages in an a-didactical situation to put the student in the position of being a simple actor in an a-didactical situation.

Elements of Didactical Situations in the Lesson

For tracing the features of TDS in the lesson, let us analyse Stage 3 (group counting and sharing: addition with two-digit numbers and one-digit number with regrouping ones). It followed the stage during which children got acquainted with a scheme describing adding two numbers without regrouping (Fig.).

The objective of Stage 3 was represented by adding 24 + 9. The expected activity was explained by the teacher on the interactive board in front of the classroom. Tens were represented by a box with ten rounds in a rectangle. Therefore 24 and 9 occurred as thc scheme on Fig. (devolution).

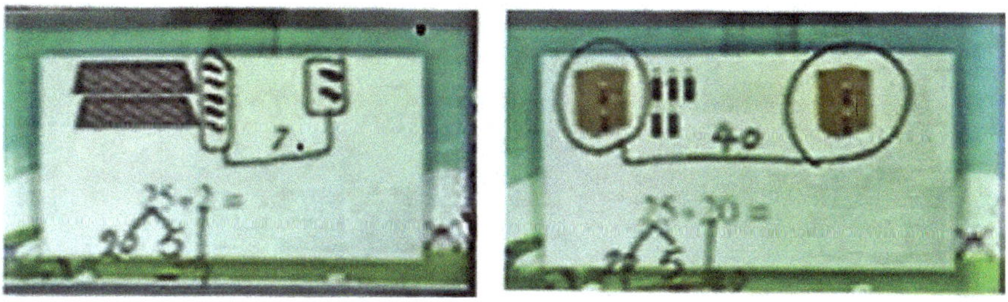

Fig. Representations on the whiteboard

Fig. Representation of 24 and 9

Fig. Candies in boxes

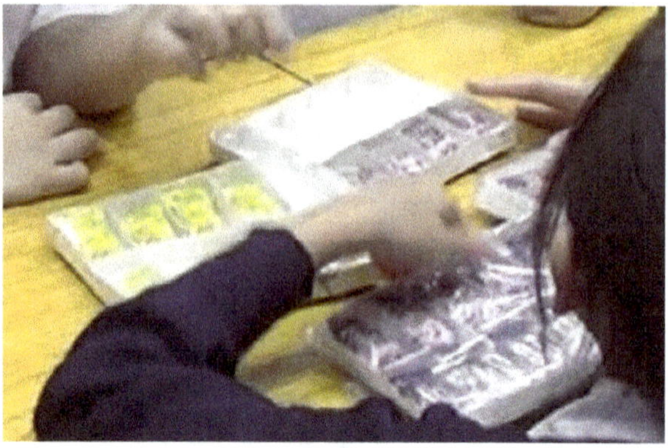

Fig. Different representations of 24 + 9

Students worked in groups. They were provided with boxes for ten candies. Twenty-four violet candies were placed in two full boxes and the third box con-tained four candies and six empty slots (Fig.). In another box, nine yellow candies were stored. When looking for the sum, students tried to add more candies in one box that was not full to have complete it to ten candies (of candies of both colours). Thus, there are three valid procedures to deal with the boxes (see Fig.). The teacher circulated through the classroom and recorded procedures applied by individual groups. From time to time, she interfered in the work of a group. By act-ing, students built the knowing of regrouping for the addition of 24 + 9. The a-didac-tical part of the situation is rather limited and does not offer too much freedom for students to act in relationship with the milieu (even filling in the boxes is already predetermined by providing groups with the given number of boxes and placing candies in a certain way in them).

The formulation and validation of the developed knowing were organised together as a whole class activity managed by the teacher. She asked representatives of groups to come to the front and model their procedure with the boxes of candies they used. At the same time, she projected a schematic description of the presented procedure and added the symbolic record of the solving strategy under the boxes fastened on the board (Fig.).

Institutionalisation was realised in the form of solving similar problems. This time, students worked in their exercise books. Individual students presented then their calculations (projected their record and explained it). The teacher did not finish with the presentation of one procedure but elicited the occurrence of others. She pushed children to work as fast as possible (the speed of solving looks to be an important factor in Chinese classrooms). The teacher insisted on students' independent description and explanation of each step of their solving procedure.

The process from the work with models to the 'abstract' way of solving the problem using mathematical symbols showed to be smooth and understandable for students.

Classroom Assessment in Whole Number Arithmetic

Mathematics teachers need to assess their students' learning progress in whole number arithmetic to be able to provide pertinent guidance during the learning process. Not only is this knowledge necessary, it is even impossible to teach without it, because teachers' teaching should build on and link to what the students already know. In other words, teachers need to have insight into students' solution strategies reflecting their mathematical thinking (Gearhart and Saxe 2005). The guidance teachers provide in their mathematics classes can be more or less effective for stimulating students' understanding, depending on whether their instruction is attuned to students' needs and possibilities for further development (cf. Butterworth 2015: '[g]etting the correct assessment is fundamental for selecting the appropriate intervention' (p. 28)). In a continual striving for providing the best possible explanations to students, teachers need to know at practically every particular moment in their classes of every single student where they are in their understanding. This echoes Schoenfeld (2014) statement that, '[p]owerful instruction 'meets students where they are' and gives them opportunities to move forward' (p. 407). Teachers can acquire insight about students' whole number arithmetic abilities by qualitative and holistic assessments, such as observing students in class and giving them open-ended tasks, providing a reliable window for knowing students' progress (cf. Black 2014). Taking this formative perspective in assessment offers teachers the possibility to adequately assess students' understanding in such a way that it directly informs their teaching of whole number arithmetic. Teachers can use the results of such formative classroom assessment to make instructional decisions, for example, to decide whether they need to adapt instruction to fit students' current mathematical understanding, repeat a particular exercise or explanation or, if students have reached a satisfactory insight, continue with their previously planned further instruction. This type of assessment is completely in 'the hands of teachers' (van den Heuvel-Panhuizen and Becker 2003, p. 683).

Reflection on the Possible Assessment Fragment of the Lesson in Macao

During the lesson observed in Macao there were some instances where the teacher appeared to be assessing her students' skills; she did not, however, seem to make formative use of the information that she could gather from the students' responses. The lead-in stage of the lesson could have functioned as a formative assessment technique, in the sense that through the students' answers to the different number combinations of 10 the teacher could have found out whether the students had memorised these number facts. Subsequently, she could have used this information to decide how to proceed with her further instruction. Checking students' knowledge of the combinations of 10, as she did in this lesson, is a very insightful starting point, as this understanding/knowledge is a prerequisite underlying the more complex calculations that were planned later in this lesson, i.e. addition with regrouping (crossing 10) until 100. However, as a formative assessment technique, it could have provided only limited useful information, as merely confirmatory evidence was gathered: all the different combinations were provided, ordered from 9 + 1, 8 + 2, etc. to 2 + 8, 1 + 9 and 10 + 0. Students had to write, as fast as possible, the result of every 'calculation', i.e. 10, next to each member of the list of combinations. Evidently, this can be done in several ways, of which only one really concerns the use of previously memorised number facts. Other solution strategies are, for example, the recognition of the minor variation in the problems, and thus concluding that all answers must be 10, without calculating, or calculating all of the combinations one by one, but fairly quickly, as they are, even for Grade 1 students, relatively easy. As a fun exercise – it was a timed competition where students could register their elapsed time themselves – this activity had its merits, but it could possibly have been more informative if it were used by the teacher with a formative assessment purpose in mind. For this, to inform further instruction, it would not be necessary to neglect the memorisation and speed objectives of the calculation competition, but it would be necessary for the teacher to change her perspective from teaching or testing for the correct answer to assessing with the purpose of better adjusting her further teaching to her students' skills, level of understanding and knowledge of number facts.

Alternative Lead-in Activity by Using Classroom Assessment

An alternative task that could be used as a classroom assessment technique in this same context is the following. To find out if students really have learned the combinations of 10 and memorised them as number facts, an informative and interactive technique could be the classroom activity in which all the students have a red and a green card with which they can show their (dis)agreement with a series of statements. The statements can be 'Are these numbers together more than 10, yes (green) or no (red)? 2 and 7, 5 and 4…' and so on, to which all the students respond instantaneously by showing a red or green card (see, for experiences with such an assessment technique in third grade in the Netherlands, Veldhuis and van den Heuvel-Panhuizen 2014, and, for the effects on student achievement, Veldhuis and van den Heuvel-Panhuizen 2015). This classroom assessment technique immediately provides the teacher with an overview of students' understanding, not only about the correctness or speed of their answers, but also whether they are able to show their answers without hesitation or if they change their minds when looking at the cards their peers show (peer feedback). Also, through the interactive nature of this activity, room for a classroom discussion on strategies students used to decide on their agreement with the statements is created, as such enabling the teacher to provide feedback on these to the students. A further possibility provided by this assessment technique is the opening for further investigation, in the same way, of students' understanding or knowledge of the analogous combinations of 100 or even 1000. Through this, the teacher could become more assessment aware, meaning that she could distinguish students practising from assessing them. With only this slight adaption of the lead-in activity and the teacher using a formative assessment perspective, she could have adjusted her instruction to the students' needs even more than she already did. A different approach, focusing more on individual students, without changing the lead-in stage, is of course also possible. Such an approach was used in Australia (Gervasoni and Parish 2015) where teachers perform clinical individual assessment interviews. These interviews can provide valuable additional information on students' instructional needs in whole number arithmetic. Likewise, the *Journal of Number* (Sensevy et al. 2015), in which individual students are regularly prompted to write down what they know about the mathematics they encountered in the preceding lessons, could be used to further enlighten the teacher about her students' understanding.

Classroom Assessment and the Lesson Plan

For the teacher to use the gathered assessment information formatively, she would need to be flexible in the use of her lesson plan. When comparing the pre-established lesson plan to the lesson we observed, there appears to be a one-to-one correspondence in content, order and type of activities she proposed. Everything was very well prepared, from the concrete materials, the posters on the whiteboard, until the hidden answers to the different problems on the video-projected slides. The questions the teacher posed appeared to be mainly focused on steering the students' responses in the direction of the prepared materials. All problems the students had to solve clearly had one, and only one, correct response. This was remarkably exemplified by the way the teacher uncovered the correct responses to the exercises; the correct answers were hiding in the presentation slides, clearly predetermined and appeared not to be open for discussion. As such, the teaching and learning were deterministic, focused on getting the only right answer and not really open for different interpretations of the problems and different solutions. In South Africa, comparable teacher behaviour in the teaching of whole number arithmetic, consisting of mostly ignoring incorrect answers and rarely juxtaposing correct and incorrect answers, has also been observed (Ekdahl and Runesson 2015). Such an approach is quite reasonable for a scripted demonstration lesson, but is contrary to the idea of formative assessment in which lesson plans are necessarily adaptable, because the information the teacher collects about students' understanding is used to adjust the actual teaching following the assessment. The use of lesson plans in this observed Macao lesson reflects a recent finding in Nanjing, China (Zhao et al. 2015), in which Chinese mathematics teachers' lesson plans on division prevailed over the possible influence of newly gathered information on students' understanding. These Chinese teachers did use (formative) classroom assessment techniques, but seemed to be unable to divert from their lesson plans and as such did not use the assessment information formatively.

The Role of Textbooks Related to Teaching of WNA

Researchers have generally agreed that textbooks play a dominant and direct role in what is addressed in instruction. Robitaille and Travers (1992) noted that a great dependence upon textbooks is 'perhaps more characteristic of the teaching of mathematics than of any other subject' (p. 706). This is due to the canonical nature of the mathematics curriculum. Teachers' decisions about the selection of content and teaching strategies are often directly set by the textbooks they use (Freeman and Porter 1989; Reys et al. 2004). Therefore, textbooks are considered to determine largely the degree of students' opportunity to learn (OTL) (Schmidt et al. 1997; Tornroos 2005). This means that if textbooks implementing a specific curriculum differ students will get different OTL (Haggarty and Pepin 2002). Consequently, different student outcomes result, as confirmed by several studies which found a strong relation between the textbook used and the mathematics performance of the students (see, e.g., Tornroos 2005; Xin 2007).

There is no doubt about the influence of textbooks on teachers' practices related to the teaching of WNA in the primary school. Knowledge about approaches to teach WNA presented in textbooks in different educational systems can provide deep insights about the diverse ways of how WNA is taught. The WNA curriculum varies across educational systems. The degree of coupling of the intended curriculum presented in textbooks with the prescribed curriculum in official documents also varies across educational systems in the world. For example, in Mainland China (Ni 2015) and Singapore (Kaur 2015), textbooks and the corresponding teaching materials are the most important vehicles used to implement the nationally mandated curriculum. The development and publishing of textbooks is closely regulated and monitored by the central government, the Ministry of Education, and there are only a few officially designated publishers who are allowed to develop textbooks and teaching manuals. However, this is not the case in some countries like the Netherlands (van Zanten and van den Heuvel-Panhuizen 2014) and France (Chambris 2015), where the respective governments only prescribe the content to be taught and publishers are left to develop textbooks without any restrictions. Likewise, in Australia and Germany, the curriculum is set by the states and follows a framework agreed by all the states, and textbooks are developed by publishers without any involvement of the authorities who prescribe the curriculum (Peter-Koop et al. 2015). At times, when publishers are left to produce the books with no guidance, a mismatch may occur, as Yang (2015) found that although the national curriculum in Taiwan emphasises number sense, few activities related to number sense are found in the elementary textbooks.

Furthermore, in some educational systems, teachers do use textbooks more often than others. In most countries where education authorities are involved in the production of the textbooks, e.g. Singapore (Kaur 2015), Hong Kong SAR (Zhang et al. 2015) and China (Cao et al. 2015), teachers use textbooks in teaching the WNA curriculum in the elementary grades. In other systems, for example, Australia, there may be a variety of textbooks used in the schools, but it is also common for teachers not to use a textbook at all, but rather devise their own tasks or draw on a variety of resources, including textbooks. In Germany, the vast majority of the teachers use one of the major textbooks, available for primary schools, to teach WNA (Peter-Koop et al. 2015). The same is true for Thailand (Inprasitha 2015). Changsri (2015) and Inprasitha (2015) noted that the use of textbooks that consist of mainly routine exercises, by teachers in Thailand, may be the cause of poor performance of Thai students in mathematics.

It is inevitable that cultural and traditional perspectives are present in textbooks for the teaching of WNA. In China, the 同文算指 *Tongwen Suanzhi* (a treatise compiled by a Chinese scholar Li Zhi-zao and an Italian Jesuit Matteo Ricci) has had significant influence on the teaching and learning of arithmetic and pedagogical design of textbooks (Siu 2015). Likewise, the number line as found in textbooks appears to be a Western aid for teaching WNA, as traces of its use can be found in the early teaching practices of most Western countries (Bartolini Bussi 2015).

The teaching of WNA has evolved with time and this is clearly evident from textbooks. In France, the classical theory of numbers which was adapted in close terms in textbooks disappeared from the textbooks and teacher books for teacher education during the 1980s (Chambris 2015). In Singapore, since the 1980s, textbooks have adopted the concrete-pictorial-abstract approach for the learning of WNA. In addition, the model method – a tool for representing and visualising relationships – has been a key heuristic students use for solving whole number arithmetic (WNA) word problems (Kaur 2015).

Comparative studies of textbooks, both within and between countries, have also shed light on the depth and breadth of the WNA curriculum across the world. Zhang et al. (2015) found that the four sets of textbooks they studied in Hong Kong followed the same curriculum guide for the design of the teaching units, and therefore there were only subtle differences in the structure and sequencing of content for two-digit subtraction of numbers in the books. However, van Zanten and van den Heuvel-Panhuizen (2014) in their study of two Dutch textbook series found that the textbook series reflected divergent views on subtraction up to 100 as a mathematical topic, which were subtraction up to 100 bridging a ten and subtraction up to a 100 without bridging a ten. The research by Inprasitha (2015) and Changsri (2015) on the teaching of WNA in Thai schools, adopting the Lesson study and open approach using Japanese mathematics textbooks, show how the Japanese textbooks are influencing the curriculum materials and teaching approaches for the teaching of WNA in classrooms of Thailand. Alafaleq et al. (2015) found that generally there was a high level of uniformity in the way the comparison of whole numbers was introduced in textbooks in China, Indonesia and Saudi Arabia.

Lastly, textbooks too may have their weaknesses. At times, they use notations that give rise to erroneous conclusions. For example, Cooper (2015) noted that in Israel elementary mathematics textbooks when 25 was divided by 3, the result 8 remainder 1 was written as 8(1), that is, 25:3 = 8(1). Similarly, when 41 was divided by 5, it was written as 41:5 = 8(1). This leads to a nonsensical deduction that 25:3 = 41:5. Cooper suggests that if the notation was revised as 25:3 = 8(1:3), it would circumvent any wrong conclusions of the equivalence relationship. Textbooks also tend to treat topics as isolated units with little connection to other units (Sowder et al. 1998). Shield and Dole (2013) found that often textbooks show the algorithmic way that topics requiring proportional reasoning are addressed with little or no connection made to related topics such as decimals, ratio, proportion and percent.

The limited focus on the connections between topics inside as well as outside mathematics and the emphasis on algorithms, closed answer form and simple connections in textbooks are hot topics addressed in many documents. For example, findings in several contributions at the International Conference on Mathematics Textbook Research and Development 2014 (ICMT-2014) highlighted this problem; see, e.g. Veilande (2014) who in his comparative study of WNA problems in Latvian fifth grade mathematics textbooks and the same in Mathematics Olympiads found that generally textbook problems focused on mathematical operations and understanding, while Olympiad problems on properties and mathematical thinking. The presentation and solving of problems in textbooks has also been the main topic of the work carried out by the Nordic Network of Research on Mathematics Textbooks (Grevholm 2011).

Concluding Remarks

The teaching and assessing of WNA is a relatively large domain. It is not possible to cover all its aspects in a chapter. Therefore, it was necessary to select some aspects which may be addressed, as we have done in this chapter.

The choice of lenses used in this chapter was guided by several aspects: the main focus was formulated in the Discussion Document (this volume, Appendix 2) and represents the agreement of all the IPC members. The second strong influence was the development of the debates in Theme 4 sessions at the ICMI Study 23 Conference in Macao, the richness of which is documented in the survey of contributions at the beginning of this chapter. The third was the decision of the team of authors to develop ideas that emerged from the lesson that they observed in Macao as part of the conference. The purpose of using the lesson to contextualise the discussion in this chapter on aspects of teaching and assessing WNA was not to critique any particular approach nor use of resource of the lesson, but rather to show how the lesson may be interpreted using various theoretical and methodical frameworks.

References

Bartolini Bussi, M. G., & Sun, X. (2015). Learning from the world: The teaching and learning of whole numbers arithmetic in the ICMI study 23. In Copirelem (Ed.), *Actes du 42e Colloque International des formateurs de mathématiques chargés de la formation des maîtres* (pp. 39–51). Besancon: ARPEME.

Bartolini Bussi, M. G., Sun, X., & Ramploud, A. (2013). A dialogue between cultures about task design for primary school. In C. Margolinas (Ed.), *Task design in mathematics education. Proceedings of ICMI study 22* (pp. 549–558). UK: Oxford. Retrieved February 10, from https://hal.archives-ouvertes.fr/hal-00834054v3.

Black, P. (2014). Assessment and the aims of the curriculum: An explorer's journey. *Prospects, 44*(4), 487–501.

Bobis, J., Clarke, B., Clarke, D., Thomas, G., Wright, R., Young-Loveridge, J., & Gould, P. (2005). Supporting teachers in the development of young Children's mathematical thinking: Three large scale cases. *Mathematics Education Research Journal., 16*(3), 27–57.

Bowden, J., & Marton, F. (1998). *The university of learning*. London: Kogan Page.

Brousseau, G. (1997). *Theory of didactical situations in mathematics 1970–1990* (F. M. Cooper, N. Balacheff, R. Sutherland & V. Warfield, Trans.). Dordrecht: Kluwer Academic Publishers.

Brousseau, G., & Sarrazy, B. (2002). *Glossaire de quelques concepts de la théorie des situations didactiques en mathématiques* (V. Warfield, Trans.). Bordeaux: DAEST, Université Bordeaux 2.

Bruner, J. S. (1960). *The process of education*. Cambridge, MA: Harvard University Press.

Cai, J. (2004). Developing algebraic thinking in the earlier grades: A case study of the Chinese elementary school curriculum. *The Mathematics Educator, 8*(1), 107–130.

Clarke, D., Cheeseman, J., Gervasoni, A., Gronn, D., Horne, M., McDonough, A., Montgomery, P., Roche, A., Sullivan, P., Clarke, B., & Rowley, G. (2002). *Early numeracy research project final report*. Melbourne: Australian Catholic University.

Clements, M. A., & Ellerton, N. (1995). Assessing the effectiveness of pencil-and-paper tests for school mathematics. In B. Atweh & S. Flavel (Eds.), *Galtha: Proceedings of the 18th annual conference of the Mathematics Education Research Group of Australasia* (pp. 184–188). Darwin: Northern Territory University.

Freeman, D., & Porter, A. (1989). Do textbooks dictate the content of mathematics instruction in elementary schools? *American Educational Research Journal, 26*(3), 403–421.

Gearhart, M., & Saxe, G. B. (2005). When teachers know what students know. *Theory Into Practice, 43*(4), 304–313.

Gervasoni, A., Parish, L., Hadden, T., Turkenburg, K., Bevan, K., Livesey, C., & Croswell, M. (2011). Insights about children's understanding of 2-digit and 3-digit numbers. In J. Clark,

B. Kissane, J. Mousley, T. Spencer, & S. Thornton (Eds.), *Mathematics: Traditions and [new] practices. Proceedings of the 23rd biennial conference of the Australian Association of Mathematics Teachers and the 34th annual conference of the Mathematics Education Research Group of Australasia* (Vol. 1, pp. 315–323). Alice Springs: MERGA/AAMT.

Gervasoni, A., & Sullivan, P. (2007). Assessing and teaching children who have difficulty learning arithmetic. *Educational & Child Psychology, 24*(2), 40–53.

Grevholm, B. (2011). Network for research on mathematics textbooks in the Nordic countries. *Nordic Studies in Mathematics Education, 16*(4), 91–102.

Haggarty, L., & Pepin, B. (2002). An investigation of mathematics textbooks in England, France and Germany: Some challenges for England. *Research in Mathematics Education, 4*(1), 127–144.

Huang, R., & Li, Y. (2017). *Teaching and learning mathematics through variation. Confucian heritage meets Western theories*. Rotterdam: Sense Publishers.

Kullberg, A. (2010). *What is taught and what is learned. Professional insights gained and shared by teachers of mathematics*. Göteborg studies in Educational sciences no 293. PhD thesis, Acta Universitatis Gothoburgensis, Göteborg.

Lo, M. L., Pong, W. Y., & Chik, P. P. M. (Eds.). (2005). *For each and everyone: Catering for individual differences through learning studies*. Hong Kong: Hong Kong University Press.

Marton, F., Tsui, A. B. M., Cjik, P. P. N., Ko, P. Y., & Lo, M. L. (2004). *Classroom discourse and the space of learning*. Mahwah: Lawrence Erlbaum Associates.

Mitchell, A., & Horne, M. (2011). Listening to children's explanations of fraction pair tasks: When more than an answer and an initial explanation are needed. In J. Clark, B. Kissane, J. Mousley, T. Spencer, & S. Thornton (Eds.), *Mathematics: Traditions and [new] practices. Proceedings of the 23rd biennial conference of the Australian Association of Mathematics Teachers and the 34th annual conference of the Mathematics Education Research Group of Australasia* (Vol. 1, pp. 515–522). Alice Springs: MERGA/AAMT.

Murata, A., & Fuson, K. (2006). Teaching as assisting individual constructive paths within an interdependent class learning zone: Japanese first graders learning to add using 10. *Journal for Research in Mathematics Education, 37*(5), 421–456.

Reys, B. J., Reys, R. E., & Chávez, O. (2004). Why mathematics textbooks matter. *Educational Leadership, 61*(5), 61–66.

Robitaille, D. F., & Travers, K. J. (1992). International studies of achievement in mathematics. In D. A. Grouws (Ed.), *Handbook of research on mathematics teaching and learning* (pp. 687–709). New York/Reston: Macmillan/National Council of Teachers of Mathematics.

Runesson, U. (2005). Beyond discourse and interaction. Variation: A critical aspect for teaching and learning mathematics. *Cambridge Journal of Education, 35*(1), 69–87.

Schmidt, W., McKnight, C., & Raizen, S. (1997). *A splintered vision: An investigation of U.S. science and mathematics education*. Boston: Kluwer Academic Publishers.

Schoenfeld, A. H. (2014). What makes for powerful classrooms, and how can we support teachers in creating them? A story of research and practice, productively intertwined. *Educational Researcher, 43*(8), 404–412.

Selter, C. (1998). Building on children's mathematics: A teaching experiment in grade three. *Educational Studies in Mathematics, 36*(1), 1–27.

Shield, M., & Dole, S. (2013). Assessing the potential of mathematics textbooks to promote deep learning. *Educational Studies in Mathematics, 82*(2), 183–199.

Sowder, J., Armstrong, B., Lamon, S., Simon, M., Sowder, L., & Thompson, A. (1998). Educating teachers to teach multiplicative structures in the middle grades. *Journal of Mathematics Teacher Education, 1*(2), 127–155.

Steffe, L., Cobb, P., & von Glasersfeld, E. (1988). *Construction of arithmetical meanings and strategies*. New York: Springer-Verlag.

Sullivan, P., Askew, M., Cheeseman, J., Clarke, D., Mornane, A., Roche, A., & Walker, N. (2015). Supporting teachers in structuring mathematics lessons involving challenging tasks. *Journal of Mathematics Teacher Education, 18*(2), 123–140.

Sun, X. (2011a). An insider's perspective: "variation problems" and their cultural grounds in Chinese curriculum practice. *Journal of Mathematical Education, 4*(1), 101–114.

Sun, X. (2011b). "Variation problems" and their roles in the topic of fraction division in Chinese mathematics textbook examples. *Educational Studies in Mathematics, 76*(1), 65–85.

Sun, X. (2016). *Spiral variation: A hidden theory to interpret the logic to design mathematics curriculum and instruction in Mainland China (in Chinese).* Singapore: World Scientific Publishing Company.

Sun, X., Neto, T. B., & Ordóñez, L. E. (2013). Different features of task design associated with goals and pedagogies in Chinese and Portuguese textbooks: The case of addition and subtraction. In C. Margolinas (Ed.), *Task design in mathematics education. Proceedings of ICMI study 22* (pp. 409–418). Oxford. Retrieved February 10, from https://hal.archives-ouvertes.fr/hal-00834054v3.

Tornroos, J. (2005). Mathematics textbooks, opportunity to learn and student achievement. *Studies in Educational Evaluation, 31*(4), 315–327.

van den Heuvel-Panhuizen, M., & Becker, J. (2003). Towards a didactical model for assessment design in mathematics education. In A. J. Bishop, M. A. Clements, C. Keitel, J. Kilpatrick, & F. K. S. Leung (Eds.), *Second international handbook of mathematics education* (pp. 689–716). Dordrecht: Kluwer Academic Publishers.

van den Heuvel-Panhuizen, M., & Drijvers, P. (2014). Realistic mathematics education. In S. Lerman (Ed.), *Encyclopedia of mathematics education* (pp. 521–525). Dordrecht/Heidelberg/New York/London: Springer.

van Zanten, M., & van den Heuvel-Panhuizen, M. (2014). Freedom of design: The multiple faces of subtraction in Dutch primary school textbooks. In Y. Li & G. Lappan (Eds.), *Mathematics curriculum in school education, advances in mathematics education* (pp. 231–259). Dordrecht/Heidelberg/New York/London: Springer.

Veilande, I. (2014). Contemporary study of 5th grade textbooks: Tasks on whole numbers and their compliance with mathematics Olympiad content. In K. Jones, C. Bokhove, G. Howson, & L. Fan (Eds.), *Proceedings of the International Conference on Mathematics Textbook Research and Development 2014 ICMT-2014* (pp. 477–482). Southampton: Southampton Education School, University of Southampton.

Veldhuis, M., & van den Heuvel-Panhuizen, M. (2014). Exploring the feasibility and effectiveness of assessment techniques to improve student learning in primary mathematics education. In C. Nicol, S. Oesterle, P. Liljedahl, & D. Allan (Eds.), *Proceedings of the joint meeting of PME 38 and PME-NA 36* (Vol. 5, pp. 329–336). Vancouver: PME.

Veldhuis, M., & van den Heuvel-Panhuizen, M. (2015). *Supporting primary school teachers' assessment practice in mathematics: Effects on students' learning.* https://www.ris.uu.nl/ws/files/11827943/Veldhuis.pdf. Manuscript submitted for publication.

Vygotsky, L. S. (1980). *Mind in society*. Cambridge, MA: MIT Press.

Wang, T., Cai, J., & Hwang, S. (2015). Achieving coherence in the mathematics classroom: Toward a framework for examining instructional coherence. In L. Fang, N. Y. Wong, J. Cai, & S. Li (Eds.), *How Chinese teach mathematics. Perspectives from insiders* (pp. 111–148). Singapore: World Scientific.

Watson, A., & Mason, J. (2005). *Mathematics as a constructive activity: Learners generating examples*. Mahwah: Lawrence Erlbaum Associates.

Watson, A., & Mason, J. (2006a). Seeing an exercise as a single mathematical object: Using variation to structure sense-making. *Mathematical Thinking and Learning, 8*(2), 91–111.

Watson, A., & Mason, J. (2006b). Variation and mathematical structure. *Mathematics Teaching, 194*, 3–5.

Wright, R., Martland, J., & Stafford, A. (2000). *Early numeracy: Assessment for teaching and intervention*. London: Paul Chapman Publishing.

Xin, Y. P. (2007). Word problem solving tasks in textbooks and their relation to student performance. *Journal of Educational Research, 100*(6), 347–359.

Zazkis, R., & Leikin, R. (2007). Generating examples: From pedagogical tool to a research tool. *For the Learning of Mathematics, 27*(2), 15–21.

Cited papers from Sun, X., Kaur, B., & Novotna, J. (Eds.). (2015). Conference proceedings of the ICMI study 23: Primary mathematics study on whole numbers. Retrieved February 10, 2016, from www.umac.mo/fed/ICMI23/doc/Proceedings_ICMI_STUDY_23_final.pdf

Alafaleq, M., Mailizar, L., Wang, Y., & Fan, L. (2015). How equality and inequality of whole numbers are introduced in China, Indonesia and Saudi Arabia primary mathematics textbooks (pp. 392–398).

Askew, M., (2015). Seeing through place value: An example of connectionist teaching (pp. 399–406).

Barry, A., Novotná, J., & Sarrazy, B. (2015). Experience and didactical knowledge: The case of didactical variability in solving problems (pp. 407–414).

Bartolini Bussi, M.G. (2015). The number line: A "western" teaching aid (pp. 298–306).

Brombacher, A. (2015). National intervention research activity for early grade mathematics in Jordan (pp. 415–422).

Butterworth, B. (2015). Low numeracy: From brain to education (pp. 21–33).

Cao, Y., Li, X. & Zuo, H. (2015). Characteristics of multiplication teaching of whole numbers in China: The application of the nine times table (pp. 423–430).

Chambris, C. (2015). Mathematical foundations for place value throughout one century of teaching in France (pp. 52–59).

Changsri, N. (2015). First grade students' mathematical ideas of addition in the context of lesson study and open approach (pp. 60–67).

Cooper, J. (2015). Combining mathematical and educational perspectives in professional development (pp. 68–75).

Ekdahl, A. L., & Runesson, U. (2015). Teachers' responses to incorrect answers on missing number problems in South Africa (pp. 431–439).

Gervasoni, A., & Parish, L. (2015). Insights and implications about the whole number knowledge of grade 1 to grade 4 children (pp. 440–447).

Inprasitha, M. (2015). An open approach incorporating lesson study: An innovation for teaching whole number arithmetic (pp. 315–322).

Kaur, B. (2015). The model method: A tool for representing and visualising relationships (pp. 448–455).

Lin, P.-J. (2015). Teaching the structure of standard algorithm of multiplication with 2-digit multipliers via conjecturing (pp. 456–463).

Ni, Y. (2015). How the Chinese methods produce arithmetic proficiency in children (pp. 339–345).

Pearn, C. (2015). Same year, same school, same curriculum: Different mathematics results (pp. 464–471).

Peter-Koop, A., Kollhoff, S., Gervasoni, A., & Parish, L. (2015). Comparing the development of Australian and German children's whole number knowledge (pp. 346–353).

Sensevy, G., Quilio, S., & Mercier, A. (2015). Arithmetic and comprehension at primary school (pp. 472–479).

Siu, M-K. (2015). Pedagogical lessons from *Tongwen Suanzhi*: Transmission of *Bisuan* (Written calculation) in China (pp. 132–139).

Wong, I. N., Jiang, C., Cheung, K.C., & Sun, X. (2015). Primary mathematics education in Macao: Fifteen years of experiences after 1999 handover from Portugal to Mainland China (pp. 480–487).

Yang, D.-C. (2015). Performance of fourth graders in judging reasonableness of computational results for whole numbers (pp. 268–275).

Zhang, Q-P., Cheung, K-C., & Cheung, K-F (2015). An analysis of two-digit numbers subtraction in Hong Kong primary mathematics textbooks (pp. 488–495).

Zhao, X., van den Heuvel-Panhuizen, M., & Veldhuis, M. (2015). Classroom assessment techniques to assess Chinese students' sense of division (pp. 496–503).

9

Language and Cultural Issues in the Teaching and Learning of WNA

Xu Hua Sun and Maria G. Bartolini Bussi

Introduction

Reflections on Language and Culture Before, During and After the Macao Conference

Language is an artefact used to communicate and think (see Chapter). Languages differ not only in pronunciation, vocabulary and grammar, but also the different 'cultures of speaking'. Language plays a common, key role in conveying mathematics concepts for learning and teaching and the development of mathematical thinking. The features of language can help to make numerical concepts transparent and support the understanding that occurs in learning discourse. A cross-cultural examination of languages should thus allow us to understand the linguistic support and limitations that may foster/hinder students' learning and teachers' teaching of mathematics. This study examines number naming and structure across languages and language issues related to whole number structure, arithmetic operations and key concepts, and thus has important educational implications for whole number arithmetic.

The first author was supported by Research Committee, University of Macau, Macao, China (MYRG2015-00203-FED). The opinions expressed in the article are those of the authors.

X.H. Sun (✉)
Faculty of education, University of Macau, Macao, China
e-mail: xhsun@umac.mo

M.G. Bartolini Bussi (✉)
Department of Education and Humanities, University of Modena and Reggio Emilia, Modena, Italy
e-mail: mariagiuseppina.bartolini@unimore.it

As reported in Chapter, this study aims to foster awareness of the relevance of cultural diversity in the teaching and learning of whole number arithmetic and in related studies. As stated in ICMI Study 21 (Barwell et al. 2016, p. 17), 'language and culture are closely and intimately related and cannot be separated'. Language and culture influence each other. Language is part of culture and plays an important role in it. A language not only contains a nation's cultural background, but also reflects a national view of life and way of thinking. Hence, no discussion of language issues in whole number arithmetic can be separated from cultural background.

In this chapter, we address language and cultural issues based on different examples reported by the conference participants, which can be roughly divided as follows:

- The language of whole number arithmetic in Indo-European languages
- The colonial case in Africa
- The Chinese case

We also consider some of the educational implications.

A short outline of Chinese grammar for numbers is collected for the interested reader, who may in this way become acquainted with some background of the Chinese mathematics education. This special focus on the Chinese language and culture depends on:

- The 'perfect' match between everyday Chinese arithmetic and the mathematician's arithmetic (Sun 2015).
- The very interesting organisation of Chinese curricula that has seemingly proceeded uninterrupted since classic times.
- The presence of some original strategies (e.g. variation problems) (Sun 2011, 2016).
- The performance of Chinese students and teachers (e.g. Geary et al. 1993; Ma 1999).
- The circumstance of us meeting together in China and seeing a Chinese first-grade lesson in person.

The presence of different cultural and linguistic traditions makes participants and readers aware of the choices that have been made throughout history regarding the teaching and learning of whole number arithmetic, creating an improved awareness of cultural diversity in mathematics. This diversity does not allow for the simple adoption of curricula developed by others, unless a careful process of *cultural trans-position* (see Chapter) is started. An example at the end of the chapter illustrates this point.

Some Everyday Language Issues in Number Understanding

An assumption of the universality of whole number arithmetic has been predominant for both curriculum reformers and international evaluators. The so-called Hindu-Arabic system of numerals as number signs is considered the most effective computation tool and has consequently been adopted by countries around the world during the last century. However, in this chapter, we discuss how whole number arithmetic is not culture-free, but rather deeply rooted in local languages and cultures, and present the inherent difficulty of transposition from language and culture perspectives.

Information about how words are connected with whole number arithmetic may be found in many books (e.g. Menninger 1969; Zaslavsky 1973; Ifrah 1981; Lam and Ang 2004). We do not aim to summarise what can easily be found elsewhere. Our intention is to systematically collect some of the information and reflections shared between the conference participants who represented many cultural contexts and report on some of the features of their languages/cultures that have important educational implications.

Before approaching the topic of whole number arithmetic, we examine some of the studies conducted in the linguistic field, particularly in the field of *pragmatics*, where the contrast within the same culture between figurative meaning in everyday language and literal meaning in school arithmetic language is investigated. As the examples (Bazzanella, personal communication) come from very different languages and cultures, this phenomenon seems universal despite having different features.

The roots of this phenomenon may be found in ancient ages. In *Poetics*, Aristotle himself introduces the idea of the *metaphor*, which consists of giving a thing a name that belongs to something else. Among the different examples, one concerns numbers: "'Indeed ten thousands noble things Odysseus did,' for ten thousand, which is a species of many, is here used instead of the word 'many'" (Levin 1982, p. 24).

In ancient China, 'ten thousand' (wàn, 万) was used in a figurative way, as in the proper name for 'the Great Wall' (万里长城, *wàn lǐ chángchéng*). This name literally means 'ten thousand lǐ long wall', where *lǐ* is an ancient unit of length used to highlight the immense length of the wall.[1]

In the last decades, linguistic scholars have started to investigate the use of numbers (whole numbers) in everyday language. There are several issues involved in this usage, such as *indeterminacy* and *approximation*. Indeterminacy in language is commonly resorted to for a variety of reasons and takes several different forms (Krifka 2007; Bazzanella 2011). In some applications of numerals, exact numbers are used systematically to denote indefinite quantities. Such uses are linked to certain very specific numerals, either rather low ones (2, 3, 4, 5) or high but 'round' ones (100, 1000). They are *hyperbolic* in the sense that the number indicated cannot

[1] https://en.wikipedia.org/wiki/Great_Wall_of_China

be true; moreover, several variants of one expression (100, 1000, 10,000, etc.) often coexist (Lavric 2010). Lavric (2010) collects several examples from European languages (English, French, German, Italian and Spanish) where the meaning of whole numbers is not the same as the numerals learnt in counting. Some expressions in some languages must be interpreted in an approximate way. For instance, the sentence 'vuoi *due* spaghetti?' (do you wish to have *two* spaghettis?) among Italian speakers means 'do you wish to have some spaghetti?' Hence, 'two' is not used in its cardinal meaning but means a general number of things. Round numbers (i.e. powers of ten, such as ten, a hundred, a thousand, ten thousand, a hundred thousand, a million) are also used in hyperbolic meaning: 'I have told you a thousand times that you have to be prudent'. Fractions may be used to mean a very small number ('half' also sometimes means a part when the original is divided into two parts that may not be equal), with the numerator 'one', and a high and round numerator (e.g. 'even a millionth of a second') is used to minimise or a very close numerator and denominator used to maximise (e.g. 'it is ninety-nine point nine percent certain'). Apart from European languages, in Mandarin Chinese, approximate numerical expressions are classified into two main types with different meanings: one with the discourse marker *bā* (吧) denoting the approximate quantity and the other without an explicit marker denoting the exact quantity (Ran 2010). These aspects are studied in linguistics for their effects on translations from one language to another, when literal translation is impossible. They are not usually considered in the literature on mathematics education, although they are important to the connection (continuity vs discontinuity) between everyday language and school language.

Place Value in Different School Languages and Cultures

Some Reported Difficulties in Understanding Place Value

How is number naming in daily language related to the structure of numbers in school mathematics? How do listeners in the mathematics classroom recognise numerical concepts? What are the cognitive bases of approximate uses? What are their effects on cognitive processes? We discuss these language issues related to the comprehension of place value in the following.

Place value is the most important concept in the so-called Hindu-Arabic system, as it has a long-term effect on the comprehension of number structure and calculations. It denotes the value of a digit depending on its place or position in the number. Each place has a value of ten times the place to its right. In Chinese literature, place value is emphasised as an understanding of numeration with different units (計數单位). Recording magnitude with different units in counting is called *place value* in English-speaking communities or *positional notation* in French-speaking communities. In place value, there are two inseparable principles (Houdement and Tempier 2015):

- The *positional principle*, where the position of each digit in a written number corresponds to a unit (e.g. hundreds stand in the third place)
- The *decimal principle*, where each unit is equal to ten units of the immediately lower order (e.g. one hundred = ten tens)

However, a range of studies shows that the teaching and learning of place value/ numeration units is difficult. For example, Tempier (2013) finds low percentages of success of 104 French third graders (8- to 9-year-olds) in tasks involving relations between units: '1 hundred = … tens' (48% success), '60 tens = … hundreds' (31% success) and 'in 764 ones there are …tens' (39% success). Even in the fourth and fifth grades, no more than half of the students demonstrate an understanding that the '5' in '25' represents five of the objects and the '2' the remaining 20 objects (Kamii 1986; Ross 1989).

Bartolini Bussi (2011) mentions a similar difficulty:

> When 7-year-old students are asked to write numbers, a common mistake in transcoding from number words to Hindu-Arabic numerals shows up: some students write '10,013' instead of '113' as the zeroes on the right (100) are not overwritten by tens and units. (p. 94)

It should not be surprising that these students cannot grasp multi-digit addition and subtraction. Many curricula in the West list place value as positional knowledge only. For instance, Howe (2010) offers a critique of elementary curricula in the USA:

> Place value…is treated as a vocabulary issue: ones place, tens place, hundreds place. It is described procedurally rather than conceptually.

Bass (see Chap.) uses the problem of counting a large collection to stimulate the development of grouping with multiple units, according to the concept of place value. Young-Loveridge and Bicknell (2015) advise supporting the comprehension of place value by providing meaningful multiplication and division at the same time. Place value is inherently *multiplicative* (Askew 2013; Bakker et al. 2014) and usually introduced as part of the addition and subtraction of multi-digit numbers before children have experienced meaningful multiplication and division. In Chap., we report on artefacts designed and used to overcome some difficulties in the introduction of place value. Based on the studies that have been conducted, a language perspective on place value is rare in the mathematics education field. In this chapter, we wish to reconstruct some part of the history of place value while looking at it from the language perspective.

Transparency and Regularity of Number Languages: Some European Cases

In Europe, place value was introduced in the thirteenth century through the Arabic tradition and came into conflict with previous traditions (Menninger 1969; Lam and Ang 2004). This explains why the principle of place value continues to be a specific part of school curricula (Fuson and Briars 1990). Units of hundreds and thousands are always explicit, but units of ones and tens are always implicit and often missing in spoken languages. For example, units of ones and tens are not visible in 'thirty-one'.

Examples in European languages show that many irregularities appear in their languages; they depend on the existence of more ancient representation (with non-ten bases) or on other linguistic properties where the combination of two words forces an abbreviation. Furthermore, the order of units may be different.[2]

In English, French and German, numbers have independent names up to 12, while in Italian the suffix *dici* appears with 11 and becomes a prefix with 17 (as in French). English and German are similar from 13 to 20 (with the suffix *teen* or *zehn,* meaning 10). But from 21 the order of reading units and tens in German is opposite to that in English until 99. In French there is the memory of base 20, e.g. 70 is *soixante-dix;* 80 is *quatre-vingts.* A similar yet more complex irregularity is present in Danish: the irregularity involves the number names between 10 and 20, the inver-sion of units and tens (as in German) and a memory of a base 20 system.

When expressing 76 + 83, for example, different languages hint at different words that make the column calculation more or less difficult.

English: seventy-six plus eighty-three
French: sixty-sixteen plus four-twentys-three
Italian: seventy-six plus eighty-three
Danish: three-and-a-half-twenty-six plus four-twenty-three
Chinese: seven tens six plus eight tens three.

The transparency of the Chinese names is likely to foster students' understanding of place value.

Post-colonial Cases: Africa and Latin America

Zaslavsky (1973) wrote her fundamental books on African mathematical tradition to contrast the scarce (if any) references to Africa in Menninger (1969). In a later study, Verran (2001) reports on the Yoruba approach to whole number arithmetic. At the Macao Conference, there were two scholars from Northern Francophone (Nadia Azrou) and South-Eastern Anglophone Africa (Veronica Sarungi) who reported on the story of whole number arithmetic in the schools in their postcolonial regions.

[2] Funghi (2016) prepared a review of many different languages.

Algeria

Azrou (2015) reports the language situation in Algeria, where many different languages are spoken with different status: classical Arabic, Berber and French together with many different local dialects. Besides the different number words, Azrou reports the different meanings of 'digit' vs 'number'. Consider the following examples:

A- 1,2,3,…,9
B- 2781
C- a series of digits to design a phone number, a car number and an address, e.g. the contact number for ICMI-49 30 20 37 24 30…
In English, A are called digits, B numbers and C numbers.
In Arabic, A are called رقم *raqm* (digit) or أرقام *arqam* (digits, the plural).
In French, A are called chiffres, B nombres and C numéro.
In Berber (Tamazight is one of the oldest languages of humanity), only one word (*numro*) is used for everything.

The relationship between the French dialect and Berber language (languages used in everyday life) presents a problem. The dialect and Berber language have kept one word (*numro*) to express everything. This is a problem for students, who confuse the mathematical concepts they learn at school (for both Arabic and French) with the street mathematics used in Berber in everyday life.

The Guatemalan Case

In Guatemala, the official language is Spanish, and the indigenous population comprises 41% of the total population. There are 25 linguistic communities grouped in 4 'pueblos' (different groups of people), i.e. Ladino, Maya, Grifúna and Xinka, each with a unique identity, culture and language. Mayans comprise 81% of the indigenous population and have four linguistic communities. The formal recognition of the complex ethnic composition of this country was made in 1996 through the 'Agreement of Peace', which recognised people's right to their cultural identities. As a consequence, the Ministry of Education set up a bilingual programme in which the teaching must be done bilingually, respecting the culture and values of the indigenous people. In 2005, there were 3800 bilingual schools, and many of their teachers could speak the indigenous language but could not write or read it. Because of the characteristics of this country, primary schools work with two number systems: vigesimal (base 20) for Mayan mathematics and decimal. Numbers are read and written according to the two systems and various languages. The Mayan numeration system uses three symbols: the dot to represent a unit (●), the bar to represent five units (▬) and a third symbol to represent the zero, also called a shell or cocoa bean (⬭). With the combination of these three symbols, the first 19 numbers are written using three rules. First, from one to four, points are combined. Second, five points form a bar. Third, bars are combined up to three.

Tanzania and Other East African Countries

Sarungi (personal communication) reports on the complex situation in the part of Africa colonised by the British Empire (East African countries). The diversity in learners' first languages makes teaching mathematics in those languages difficult. For example, Tanzania has over 120 ethnic tribes with their own languages, although these belong to major language groups such as Bantu, Nilotic and Cushitic. At the same time, Kiswahili, which is a mixture of Bantu, Arabic and other languages such as Portuguese and English, has become the first language of tribes along the coast and islands of Zanzibar. In fact, Kiswahili is the national language of Tanzania and Kenya and is widely spoken in other East and Central African countries such as Burundi, Rwanda, Uganda and Democratic Republic of Congo.

In Tanzania, the language policy is to use Kiswahili as the medium of instruction in pre-primary and primary education (MOEVT 2014), even though Kiswahili is not the first language of many children, especially those in rural areas (Halai and Karuka 2013) and is learnt formally when entering school. In Uganda, the policy is to use ethnic or local languages in the first 3 years of primary school, although English is used in settings in which learners have diverse local languages (National Curriculum Development Centre n.d.). Research conducted in African contexts has pointed to the challenges of using a language that is not easily accessible to learners and even teachers in some cases, while the use of first languages in mathematics classrooms has been shown to foster more interactions between learners and teachers (Sepeng 2014).

Apart from the benefits of increased participation, the names of numbers in ethnic languages usually point to a base 10 structure (see Funghi 2016). Most African languages have a similar structure for numbers between 10 and 20, namely, 'ten' and '*digit*', where *digit* stands for a number from one to nine inclusive. Moreover, the decades from 20 to 90 have a logical structure. The wording constitutes either 'tens *digit*' to signify how many tens are taken or 'decade *digit*', such as in Simbiti. Thus, a number like 34 in African ethnic language is literally formed as 'tens three and four' or mathematically 'three tens and four'. Many children encounter the names of numbers to around 30 in a non-formal way by the time they start attending school. Thus, learning whole numbers in such ethnic languages could help learners to make sense of the structure of the numbers. However, there are challenges in taking advantage of these affordances. First, teachers may not be equipped to assist learners, due to unfamiliarity with the local language and its mathematical register (Chauma 2012). Moreover, the language policy may not be favourable to promoting the use of ethnic languages, as is the case for Tanzania, where Kiswahili is encouraged for purposes of national unity.

When widely spoken in a community, Kiswahili can bridge the gap between the multiplicities of languages, although its use for learning whole numbers can be a potential source of confusion for children, even if it is their first language. This is due to the origins of number names from Arabic and Bantu words, which results in an inconsistency in the naming of decades (for a comparison between Awahili and Arabic, see Funghi 2016).

For Bantu speakers, numbers from 1 to 20 present little problem, except for the names of 6, 7 and 9. However, non-Bantu speakers have to learn most of the names, although the structure from 11 to 20 is familiar. For most learners, there is an additional cognitive demand in learning the names of decades, which no longer adhere to the structure of Bantu or other ethnic languages but instead borrow words from Arabic. For example, there is very little link between the names for 30 and 3. In effect, children are required to learn new names for 20, 30, 40 and 50 in Kiswahili. It is only 60, 70 and 90 that have some link to 6, 7 and 9, respectively. Thus, asking children to write down a given two-digit number in words can result in confusion if, for instance, the child needs to remember the name for 30 (*thelathini*) and cannot infer it from its name, which is linked to its value (three tens). On a related point, the use of English in private pre-primary schools in Tanzania further complicates the matter, as the structure of numbers from 11 to 19 does not follow the known structure of Bantu and Kiswahili. Ultimately, even in contexts in which both learners and teachers speak the same language as the language of instruction, it is important to take into account the features of the common language that hinder or promote the learning of whole numbers in early years of schooling.

Towards Transparency: The Chinese Approach

Chinese young children perform better at facets of basic arithmetic, such as generating cardinal and ordinal number names (Miller et al. 2000), understanding the base 10 system and the concept of place value (Fuson and Kwon 1992), using decompositions as their primary backup strategy to solve simple addition problems (Geary et al. 1993) and calculation (Cai 1998). A comparative study (Geary et al. 1992) indicates that the addition calculating scores of Chinese students is three times that of American students. Specifically, Chinese students use more advanced strategies and exhibit faster retrieval speeds. American students use counting strategies (e.g. counting fingers or verbal counting) more frequently than their Chinese counterparts. Chinese students use retrieval strategies more frequently than their American counterparts (He 2015). However, most studies have provided various explanations for these findings, such as parents' high expectations for education, the diligence of the students and the effectiveness of the teachers. Ni (2015) argues that elementary school curricula, textbooks, classroom instruction and the cultural values related to learning mathematics have contributed to the arithmetic proficiency of Chinese children and the establishment of arithmetic as a social-cultural system.

The 2013 PISA results in mathematics (from the test taken in 2012) showed that the highest performers were located in Asian countries, placing in the following order: (1) Shanghai (China), (2) Singapore, (3) Hong Kong (China), (4) Taiwan, (5) South Korea, (6) Macao (China) and (7) Japan. All of these countries have used languages that share the same ancient Chinese number tradition.

Some authors have studied the Chinese language and culture in mathematics education in the last few decades. For instance, ICMI Study 13 (Leung, Graf and

Lopez-Real 2006) first focused on a comparison of East Asia and the West. It was followed by a trend of studies and volumes about Chinese tradition in mathematics education (Fan et al. 2004, 2015; Li and Huang 2013; Wang 2013). The specific issue of language has been addressed by many authors such as Galligan (2001) and Ng and Rao (2010), and other authors such as Fuson and Li (2009) and Xie and Carspecken (2007) have compared educational materials in China and the USA.

This phenomenon relates to a large number of teachers and students. In China, there are nearly 2.63 hundred million primary school students. Moreover, the ancient Chinese literature affected the development of mathematics in most East Asian countries (e.g. Japan, Korea, Vietnam) (Lam and Ang 2004) in terms of the convention of place value.

The Chinese Approach to Arithmetic

The Ancient History

The Chinese approach to numerals in primary schools shows consistency among the features of Chinese language, the names of numbers and the use of artefacts for representing numbers and computing, which can be traced back to the tradition of teaching numbers in China in fourteenth century BCE (Guo 2010). The long tradition is reflected in a range of ancient Chinese arithmetic works, such as the official mathematical texts for imperial examinations in mathematics used a thousand years ago:

The *Suàn shù shū, Writings on Reckoning* (算数书) (202–186 BCE)
Zhoubi Suanjing (周髀算经) (100 BCE)
The Nine Chapters on the Mathematical Art (九章算术) (100 BCE)
The Sea Island Mathematical Manual (海岛算经) (about 225–295 CE)
The Mathematical Classic of Sun Zi (孙子算经) (500 CE)
The Mathematical Classic of Zhang Qiujian (张丘建算经) (500 CE)
Computational Canon of the Five Administrative Sections (五曹算经) (1212 CE)
Xia Houyang's Computational Canons (夏侯阳算经) (1084 CE)
Computational Prescriptions of the Five Classics (五经算术)
Jigu Suanjing (缉古算经) (625 CE)
Zuisu (缀术) (500 CE)
Shushu jiyi (数术记遗) (about 200 CE)

In this section, we offer a short outline of Lam and Ang's (2004) *Fleeting Footsteps*, a long history drawing on an important reference.

In the general history of numbers, the importance of the Chinese tradition is not always acknowledged. For instance, Ifrah (1981) claims that place value is an Indian invention. Dauben (2002) strictly criticises this error:

> One Chinese source of which Ifrah is apparently unaware is the *Sun Zi Suanjing* 孙子算经 (The Mathematical Classic of Sun Zi), written around 400 CE. (p. 37)

This text has been available in an English translation since 1992 in *Fleeting Footsteps,* an edition prepared with extensive commentary by Lam and Ang, who later published a more extended edition (Lam and Ang 2004). This source not only gives a complete description of Chinese rod numerals, but also describes in detail ancient procedures for arithmetic operations. The most ambitious part of Lam and Ang's study argues that the Hindu-Arabic number system had its origins in the rod numeral system of the Chinese. The most persuasive evidence Lam and Ang offer is the fact that the complicated, step-by-step procedures for carrying out multiplication and division are identical to the earliest but later methods of performing multiplication and division in the West using Hindu-Arabic numerals, as described in the Arabic texts of al-Khwārizmī, al-Uqlīdisī and Kūshyār ibn Labbān (see the extensive review in Lam and Ang 2004). Guo (2010) explains that the Chinese system was transmitted to India during the fifth to ninth centuries, to the Arabic empire in the tenth century and then to Europe in the thirteenth century through the Silk Road. In 1853, Alexander Wylie, Christian missionary to China, refuted the notion that 'the Chinese numbers were written in words at length' and stated that in ancient China calculation was carried out by means of counting rods and that 'the written character is evidently a rude presentation of these', showing both the arithmetic procedure and the decimal place value notation in their numeral system through the use of rods. Wylie believed that this arithmetic method invented by the ancient Chinese played a vital role in the advancement of all fields that required calculations. After being introduced to the rod numerals, he wrote:

> Having thus obtained a simple but effective system of figures, we find the Chinese in actual use of a method of notation depending on the theory of local value [i.e. place value], several centuries before such theory was understood in Europe, and while yet the science of numbers had scarcely dawned among the Arabs. (p. 85)

In a review of the first edition of the *Archives internationales d'histoire des sciences*, Volkov (1996) writes that the book 'may provoke a strong reaction from historians of European mathematics'. Nevertheless, Volkov emphasises one of the book's great strengths:

> The emphasis made by the authors on the great importance of studying Chinese methods of instrumental calculators as well as numerical and algorithmic aspects of Chinese mathematics, which otherwise cannot be understood properly. (p. 158)

Chemla (1998) suggests adopting a prudent attitude towards this controversy:

> The *nine chapters* share with the earliest extant Indian mathematical writing (6th c.) basic common knowledge, among which is the use of a place-value decimal numeration system. Such evidence allows no conclusion as to where this knowledge originated, a question which the state of the remaining sources may prevent us forever answering. Instead, it suggests that, from early on, communities practicing mathematics in both areas must have established substantial communication. (p. 793)

This historic origin could be helpful for understanding why the Chinese (Eastern individuals) have found it so easy to grasp this concept, why it is so late to develop in Europe, how number heritage has been shaped and how we can advise on the number practices or tools used to strengthen the comprehension of place value. In the following, we elaborate on the Chinese approach to arithmetic as representative of East Asia. We begin by considering some elements of the Chinese approach to numbers and computation and then discuss some of the educational implications.

Although ancient Chinese mathematicians did not develop a deductive approach, they made advances in inductive algorithm and algebra development (Guo 2010). The *Zhoubi Suanjing* (周髀算經), the oldest complete surviving mathematical text compiled between 100 BCE and 100 CE, contains a statement highlighting the analogy nature of Chinese tradition:

> In relation to numbers, you are not as yet able to generalize categories. This shows there are things your knowledge does not extend to, and there are things that are beyond the capacity of your spirit. Now in the methods of the Way [that I teach], illuminating knowledge of categories [is shown] when words are simple but their application is wide-ranging. When you ask about one category and are thus able to comprehend a myriad matters, I call that understanding the Dao. … This is because a person gains knowledge by analogy, that is, after understanding a particular line of argument they can infer various kinds of similar reasoning …Whoever can draw inferences about other cases from one instance can generalize. (Quoted in Cullen 1996, pp. 175-176)

Since antiquity, the major focus of Chinese mathematics has been on numbers and computations as collections of prescriptions similar to modern algorithms. Mathematics is called *shùxué* 数学 ("shù" meaning 'number') in Chinese. Knotted cords and tallies (see Section) were mentioned in ancient Chinese literature (Martzloff 1997, p. 179), following multiplicative-additive rules. The Chinese used bamboo rods to count (see the information about counting rods in Section), and this activity fostered the creation of a systematic way to represent numbers. The first nine numerals formed by the rods are presented in Fig.

According to Lam and Ang (2004), the number presentation principle was initially introduced as follows:

> Numerals in tens, hundreds and thousands were placed side by side, with adjacent digits rotated, to tell each apart. For example, 1 was represented by a vertical rod, but 10 was represented by a horizontal one, 100 by a vertical one, 1000 by a horizontal one and so forth. Zero was represented by a blank space so the numerals 84,167 and 80,167 would be as shown [see Figure]. … Although most books on the early history of mathematics, especially the recent ones, have mentioned the Chinese rod numerals, they have failed to draw attention to a very important fact that the ancient Chinese had invented a positional NOTATION. Any number, however large, could be expressed through this place value notation which only required the knowledge of nine signs. I should add that in the current

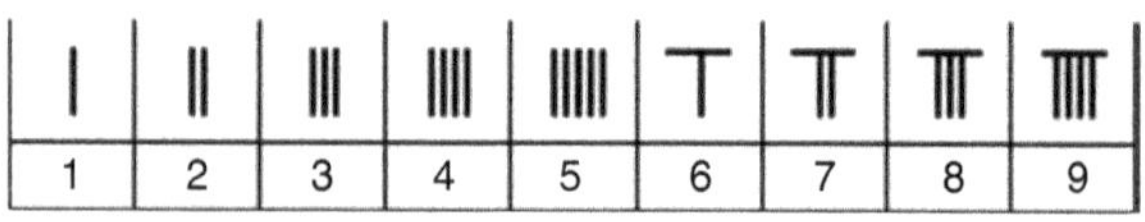

Fig. The Chinese Rod representation of the first nine numerals

Fig. Rod representation of multidigit numbers

Fig. An ancient drawing of a *suàn pán* (算盘) with measurement units (https://commons.wikimedia.org/wiki/File:Ming_suanpan.JPG)

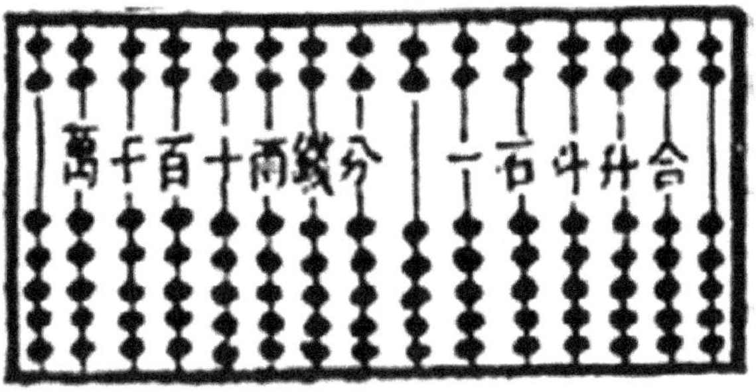

> more sophisticated written form, a tenth sign, in the form of zero, is required (Lam and Ang 2004, p. 1)

The translation of the computation principle was initially introduced as follows:

> In the common method of computation [with rods] (*fán suàn zhĭ fă*, 凡算之法), one must first know the positions (*wèi*, 位) [of the rod numerals]. The units are vertical and the tens horizontal, the hundreds stand and the thousands prostrate; thousands and tens look alike and so do ten thousands and hundred. (Lam and Ang 2004, p. 193)

A feature of Chinese mathematics is the ancient use of the counting rods (算筹, *suàn chóu*) on a table (counting board, *jìshù băn*, 计数板, Fig.). The counting board was used to make computations (arithmetic operations, extracting roots) and solve equations. The rules for using the counting board are carefully described by Lam and Ang (2004), who highlight the feature of introducing procedures in pairs: the procedure for subtraction is the inverse of that for addition, and the procedure for division is the inverse of that for multiplication. Chemla (1996) highlights that the position in the counting board is stable (see Section, which reports on the wording of the elements of arithmetic operations).

Chinese Language Foundation to Place Value

The concept of place value is dominantly used in counting rod or *suàn pán* (算盘) and written numerals (Sun 2015). Moreover, place value can be traced to the use of base 10 and conversion rates for measurements, classifier grammar and the part-part-whole structure with radicals and characters in local language. This language origin can be helpful in understanding why Western students find it so difficult to grasp the concept of place value and why it has developed so late in Europe from a language perspective.

Base 10 and the Conversion Rate for Measurement

The Chinese system had a base 10 convention for representing quantities (Lam and Ang 2004; Martzloff 1997; Sun 2015). This was consistent with the conversion rate between the measurement units of length and volume, since the first emperor *(qin shi huang* 秦始皇) who unified the whole of China in third century BCE introduced a metric system for measurements.[3] Except for weight units, units of length and volume had base 10 conversion rates. For example, the conversion rate of length units was expressed as follows (Lam and Ang 2004):

1 *yin*(引) = 10 *zhang* (丈) = 100 *chi* (尺) = 1000 *cun* (寸) = 10,000 *fen* (分) = 100,000 *li* (釐) = 1,000,000 *hao* (毫).

The conversion rate of weight units was:

1 *liang* (两) = 10 *qian* (钱) = 100 *fen* (分) = 1000 *li* (釐) = 10,000 *hao* (毫) = 100,000 *si* (丝).

The conversion rate of volume units was:

1 *gong* (斛) = 10 *dou* (斗); 1 *dou* (斗) = 10 *sheng* (升).

The ancient conversion rate of time units was 100 before the Western Zhou dynasty:

1 *shi* (时) = 100 *ke* (刻); 1 *night* (晝夜) = 5 *geng* (更).

The first money conversion rate was 10 in ancient China:

1 *peng* (朋) = 10 *ke* (貝).

Besides the measurement unit systems, the Chinese system had a base 10 convention for representing numerals using number characters and corresponding number units (Zou 2015). This can be ascribed to the Yellow Emperor in the sixth century book by Zhen Luan, *Wujing suanshu* (五經算術 Arithmetic in Five Classics) (Guo 2010). The first five number units, i.e. ge (個), shí (十), *bǎi* (百), *qiān* (千) and *wàn* (萬), always represent 1, 10, 10^2, 10^3 and 10^4, respectively. The other number units vary with different systems of number notation.

Shushu jiyi 《数术记遗》 written by XuYue (徐嶽) during the Eastern Han dynasty (50–200 CE) recorded the early number naming principle: the conversion rate of the down number (*xiashu* 下数), i.e. the standard number, was 10; the conversion rate of the middle number (*zhongshu* 中数), i.e. the large number, was 10,000; and the conversion rate of the up number (*shangshu* 上数), i.e. the largest number, was the square of the number unit.

Looking at decimal and fraction numbers, the following spoken numeration units were used to denote small orders of magnitude in Sunzi Suanjing (Lam and Ang 2004) in ancient China. The negative power of 10 was stressed in daily spoken numerals: 10^{-4}絲 *si*, 10^{-3} 毫 *hao*, 10^{-2} 釐 *li* and 10^{-1}分 *fen*.

[3] It may be interesting to compare the situation in China with the situation in Europe, where the metric system was introduced at the end of the eighteenth century, and the situation in the USA and UK, where metricisation is still controversial. For example, the conversion rates of inches, feet, yards and miles are non-10: 12 inches = 1 foot, 3 feet = 1 yard, 5280 feet = 1 mile. One US fluid ounce is 1/16 of a US pint, 1/32 of a US quart and 1/128 of a US gallon.

Classifiers

All number units in the Chinese language are called *classifiers* (*Liàngcí* 量詞). In English, it is natural to use measurement words to describe the quantity of a continuous noun (i.e. to identify a specific unit to make the quantity countable). For example, in 1 m of cloth, 1 ml of water and 1 kg of meat, the measurement units of m, ml and kg are, respectively, required. However, it is natural not to use measurement words to describe the quantity of countable nouns (e.g. one apple, five ducks and three desks). There are hundreds of different classifiers, all of which reflect the objects to be counted. In Chinese, both uncountable and countable nouns need measurement words known as classifiers. Consider one *ge* (個) apple, five *zhi* (只) ducks and three *zhang* (張) desks, in which *ge* ("unit of fruit"), *zhi* ("unit of animal") and *zhang* ("unit of object") play the role of measurement word as units. This is a kind of Chinese grammar used to describe quantity (數量), which requires numbers and classifiers. The classifiers are called number units (Zou 2015), numeration units (Houdement and Tempier 2015), number ranks (Lam and Ang 2004) or number markers (Martzloff 1997).

The column units left of the *suàn pán* (算盘) shown in Fig. read from left to right as follows:

Wan (萬 ten thousand), *qian* (千 thousand), *bai* (百 hundred), *shi* (十 ten), *liang* (兩 weight unit 1 *liang* = 1/16 jin), 錢 *qian* (10 *qian* = 1 *liang*), 分 *fen* (weight unit, 10 *fen* = 1 *qian*).

The column units right of the *suàn pán* shown in Fig. read as follows from left to right:

一 ones, 石 *shi* (volume unit, 10 *dou* = 1 *shi*); 斗 *dou*, 升 *sheng*, 合 (1 斛 = 10 斗, 1 斗 = 10 升, 1 升 = 10 合).

The units are used as column units. Weight, volume and numeration units have the same position in a functioning calculation. This indicates that numeration units have the same role as that of measurement units in Chinese (Martzloff 1997).

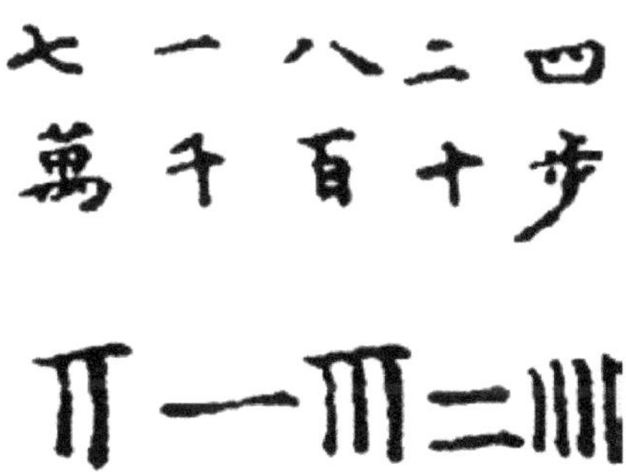

Fig. The number 71,824, written by the mathematician Jia Xian during the Song dynasty (960–1279)

一个、二个、……;　　一十一个、一十二个、……;　　二十个，二十一个....

Fig. The oral counting in Chinese. One ones, two ones …; ten ones, one ten and one ones, one ten and two ones, …; two ten ones, two tens and one ones …

Following Allan (1977), there are about 50 languages in the world with this feature, some in the Far East and some in other parts of the world. We discuss the case of the Chinese language, where numeral classifiers are systematically used, in detail as follows. Following Senft (2000), numeral classifiers are defined in the following way:

> In counting inanimate as well as animate referents the numerals (obligatorily) concatenate with a certain morpheme, which is the so called 'classifier'. This morpheme classifies and quantifies the respective nominal referent according to semantic criteria. (p. 15)

There are many classifiers in Chinese, as each type of counted object has a particular classifier associated with it. This is a weak rule, as it is often acceptable to use the generic classifier (*gè,* 个) in place of a more specific classifier. The generic classifier (*gè,* 个) is not translated into English, but may be considered as a kind of unit (a 'one'). The generic classifier may be considered the prototype of units in place value representation.

Besides the generic classifier (*gè,* 个), other units of higher value have been introduced in Chinese to represent numbers: ten (*shi* 十), hundred (*bǎi* 百), thousand (*qiān* 千) and ten thousand (*wàn* 万) . A very interesting example from ancient Chinese is given in Figure in the *Yongle Encyclopedia* (1408).[4] In the example shown in Fig., the number 71,824 is represented to indicate the digit and number (or measurement) unit. In this case, the unit is 步 (*bù*, i.e. step, an ancient length unit).

In particular:

- The first line '七一八四二' represents the number value '71,824'.
- The second line represents the units 萬 (*wàn*, ten thousand), 千 (*qiān*, thousand), 百 (*bǎi,* hundred), 十 (*shí*, ten) and 步 (*bu*, or 'step', an ancient length unit).
- The third line represents the number using the ancient rod numerals, hinting at the counting rods discussed previously. The number units have the same position as the measurement unit (bu).

Classifiers are used also in the recitation of numerals when counting objects, so that both oral and written numerals are kept consistent with each other.

In Fig., ten-two hints at an addition procedure of 10 + 2, while two tens hints at a multiplication procedure of 2 × 10. Hence, the Hindu-Arabic number 24 is translated into the Chinese language as 'two tens and four ones' (二十四个).

The legend shows a literal translation into English (numeral and classifier). In the translation, there is ambiguity between one (number) and one (classifier or unit),

---

[4] https://en.wikipedia.org/wiki/Yongle_Encyclopedia. See also http://www.wdl.org/en/item/3019/

which in Chinese are written (一, 个) and said (yī and gè) in two different ways. The same happens for 10, which in English is both a number and a unit.

In other languages (e.g. Italian), the situation may be less ambiguous, as 'uno' and 'dieci' (numbers 1 and 10) are different from 'unità' and 'decina' (unit), but the use of terms like the latter in the reading of numbers is limited to school practice (decomposition of a given number in unit, ten, hundred and so on).

In Chinese, classifiers are also used in interrogative questions, e.g. 多少 (*duōshǎo*), which means 'How much? How many?', and the right classifier must follow. When this term is used in an arithmetic word problem, e.g. in additive problems, the same classifier is used for both the data and question. For example, if five *zhi* (只) ducks swim in a river, and then two *zhi* (只) ducks join them, how many *zhi* (只) ducks are there altogether? This example shows that *zhi* (只) must be used for both the data and question.

By identifying classifiers of quantity, concrete numbers with units of the same name (same classifiers) are defined *like numbers*. A principle for arithmetic operations with like numbers is also constructed in everyday language.

Principle of addition/subtraction: only like numbers can be directly added or subtracted. Two *zhi* (只) ducks can be added to three *zhi* (只) ducks. Two *zhi* (只) ducks cannot be added to three dozen *da* (打) or groups of ducks.

Principle of multiplication: only unlike numbers can be directly multiplied. For example, three groups of *zhi* (只) ducks swim at the river. Each group comprises four *zhi* (只) ducks. How many *zhi* (只) ducks are there in total? The answer is 4 *zhi* (只) ducks * 3 groups = 12 *zhi* (只) ducks.

Principle of division:

- *With like numbers (measure division)*: for example, 12 *zhi* (只) ducks swim in the river. Each group comprises four *zhi* (只) ducks. In this case, 12 and 4 are like numbers. How many groups are there in total? The answer is 12 *zhi* (只) ducks/4 *zhi* (只) ducks = 3 groups.
- *With unlike numbers (partitive division)*: for example, 12 *zhi* (只) ducks swim in the river. We plan to group them into three groups. Here, 12 and 3 are unlike numbers. How many *zhi* (只) ducks are in each group? The answer is 12 *zhi* (只) ducks/3 groups = 4 *zhi* (只) ducks per group.

Classifiers are one of the most important elements required in word problem-solving. Generally, Chinese curricula do not need a section to differentiate partitive division from measure division, as the grammar of classifiers is enough to introduce the distinction.

Radicals and the Part-Part-Whole Structure

Radicals (部首 *bù shǒu* 'section headers') constitute the basic writing unit. Most (80–90%) of Chinese characters are phonetic-semantic compounds, combining a semantic radical with a phonetic radical. Chinese words have a compound or

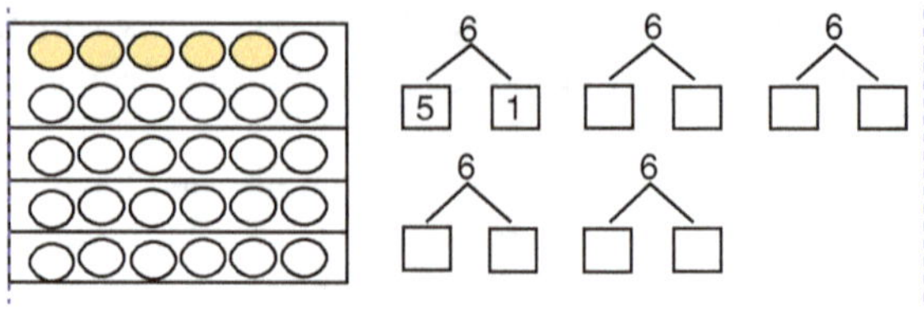

Fig. The decomposition of 6 in many different ways as 5 + 1, 4 + 2 and so on (Mathematics Textbook Developer Group for Elementary School 2005, p. 42)

part-part-whole structure. The compound can be seen in the structure of Chinese number words. For example, as shown previously, the Chinese refer to the number 12 as ‘ten-two’ rather than as a single word such as ‘twelve’.

The idea of a part-part-whole structure appears in a more general way in number computations. A number (a whole) may be conceived as the sum of two parts in dif-ferent ways (see Fig.).

This idea may be connected with the use of artefacts (either counting boards *jìshù bǎn*, 计数板 with rod numerals 算筹; *suàn chóu* or *suàn pán* 算盘). For instance, in the *suàn pán*, it is important to ‘make a ten’ by replacing two groups of five beads with one bead in the tens place if one has to calculate the following while also exploring the associative and commutative properties of addition:

$$15 + 7 = (10 + 5) + (5 + 2) = 10 + 5 + 5 + 2 = 10 + 10 + 2$$

The practice of composing/decomposing numbers is exploited to carry out very fast calculations.

Conceptual Naming of Fractions

The Nine Chapters on the Mathematical Art (九章算術; *Jiǔzhāng Suànshù*) was composed by several generations of scholars from the tenth to second century BCE, with its latest stage composed from the second century CE. According to Guo (2010), it gave the first fraction theory in the world. These are the procedures called *he fen* (addition 合分: Problems 7–9), *jian fen* (subtraction 減分: Problems 10–11), *ke fen* (comparison 課分: Problems 12–14), *ping fen* (arithmetic mean 平分: Problems 15–16), *cheng fen* (multiplication 乘分: Problems 19–25) and *jing fen* (division 經分[5]: Problems 17–18) (Sun and Sun 2012).

Martzloff (1997) observes, ‘In Chinese mathematics, by far the most common notion of fraction is that which comes from the notion of dividing a whole into an equal number of equal parts (sharing)’ (p. 192). He quotes examples such as 三分之二 (*sān fēn zhī èr*), meaning ‘two thirds’. The word 分 (*fēn*) suggests the idea of

[5] Like 徑分 in *The Nine Chapters on the Mathematical Art*, in ancient times, 徑 and 經 were regarded as the same word.

sharing, as etymologically its upper component bā (八) means 'to share', while its lower component represents a knife (刀, *dāo*). The order of reading (and writing) is denominator first and numerator second and may be literally translated as 'of three parts, one'. Martzloff (1997) continues:

> The denominator and the numerator are then respectively called *fēn mǔ* (分母 the 'mother' of the sharing) and *fēn zǐ* (分子 the 'son' of the sharing). The inventor of these expressions was thinking of a pregnant mother and her child, thus highlighting both the difference in size and the intimate link between the two terms. (p. 103)

According to Needham and Wang (1959) and Guo (2010), decimal fractions were called tiny numbers (微數 *wēi shù*), first developed and used by the Chinese in first century BCE by Liuhui (劉徽) (Guo 2010).

Arithmetic Operations

Here, we explain how addition and subtraction were introduced into Chinese tradition. The links between addition and subtraction were highlighted in the ancient textbooks. In 1274, Yang Hui observed, 'Whenever there is addition there is subtraction' (quoted in Siu 2004, p. 164).

This strict link is evident in the wording of operations. The strong regularity is evident in the following list:

加 – *jiā* – addition.
加数 – *jiā shù* – addend.
减 – *jiǎn* – subtraction.
减数 – *jiǎn shù* – subtrahend, literally 'subtracting number'.
被减数 – *bèi jiǎn shù* minuend, literally 'subtracted numbers'.
乘法 – *chéngfǎ* – multiplication.
被乘数 – *bèi chéng shù* – literally 'multiplied number'.
乘数 – *chéng shù* – literally 'multiplying number'.
除法 – *chúfǎ* – division.
被除数 – *bèi chú shù* – dividend, literally 'divided number'.
除数 – *chúshù* – divisor, literally 'dividing number'.
被 (*bèi*) is the most common word used in Chinese to create the passive verb form.

This regularity is meaningful, especially when compared with the wording in Western languages. Schwartzman (1994) points out that many English mathematics terms are borrowed from Greek and that Latin-derived terms bear no inherent meaning. For example, the English words 'minuend' and 'subtrahend', which come from Latin words and thus have little meaning today for English-speaking children in contrast with Chinese subtracted and subtracting numbers, directly embody the subtraction relationship without the exchange law. (The same is true in other Western languages.)

Addition and subtraction are carried out using counting rods (算筹 *suàn chóu*) (see Section) by simply grouping (組合 *zǔhé* making the bundle) or ungrouping

Fig. Addition in the Chinese textbook (Mathematics Textbook Developer Group for Elementary School, 2005, vol. 2, p. 62.)

Fig. Subtraction in the Chinese textbook (Mathematics Textbook Developer Group for Elementary School, 2005, vol. 2, p. 68.)

(解組 *jiě zǔ* opening the bundle) the rods (see Figs. Mathematics Textbook Developer Group for Elementary School 2005).

When the abacus (*suàn pán* 算盘) is introduced, fingering is complex (Fig.) and wording may become different (Fig.):

进一 – *jìn yī* – forward (towards the unit of higher value, e.g. when 10 units becomes a ten

退一 – *tuì yī* – backward (towards the unit of lower value, e.g. when a ten becomes 10 units

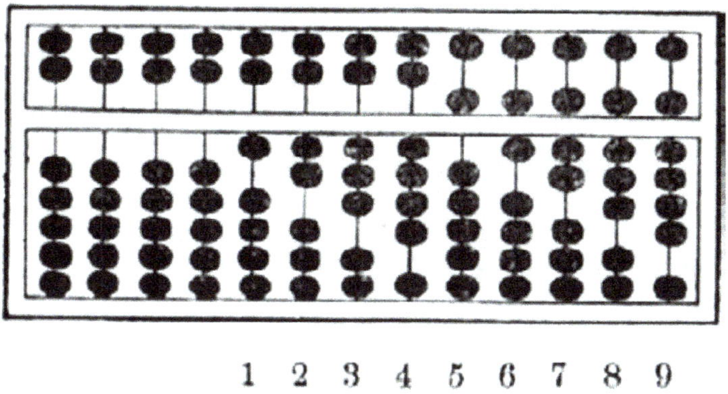

Fig. Representation of 123456789 in a Chinese *suàn pán* (Kwa 1922, p. 6)

Fig. Wording on *suàn pán*: forward and backward

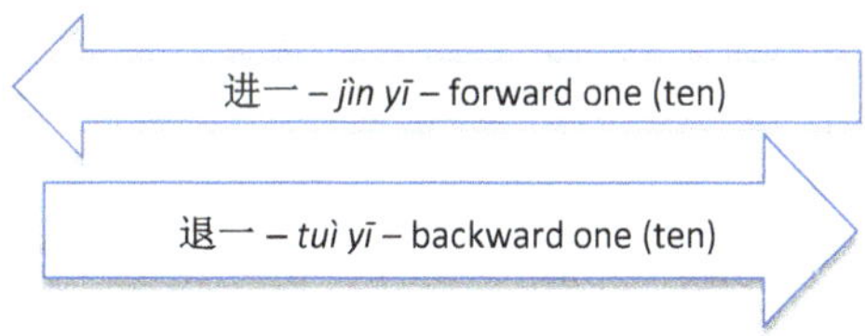

The following images are taken from Kwa (1922), an old handbook of the Chinese abacus that was included as a gift for participants at the Macao Conference.

These features are interesting, as in both cases they emphasise the inverse relation between addition and subtraction, which are described by means of inverse verbs. Division is based on multiplication, as it is the inverse of multiplication and uses a scheme that is symmetric with respect to the multiplication performed in rod calculations (adapted from Martzloff 1997, p. 217) (Table).

We analyse the differences from Western languages where this link is not highlighted as follows.

Mathematical Relational Thinking: Equality

A range of studies has advised emphasising not only numerical computation but also quantitative relationships (Ma 2015; Bass 2015; see also Chapters of this volume). The relational thinking of equality constitutes a central aspect of equations and algebra thinking (Cai and Knuth 2011). Equality is a key concept, but sometimes prob-lems are presented in Western curricula. Li et al. (2008) show that Chinese curricula introduce the equal sign in a context of relationships and interpret the sign as 'balance', 'sameness' or 'equivalence'. In the following, we review the history of the equal sign and address the approach to the relational view of the equal sign and equality.

Table The symmetric scheme in *Sunzi Suanjing*

Multiplication	Multiplication	Division	Position
Multiplicand	Multiplier	Quotient (*shang* 商)	Upper
Product	Product	Dividend (*shi* 實)	Central
Multiplier	Multiplicand	Divisor (*fa* 法)	Lower

The History of the Equal Sign '=' in Europe

The equal sign ('=') was invented (and used in its relational meaning) in 1557 by Welsh mathematician Robert Recorde (in his work *The Whetstone of Witte*), who was fed up with writing 'is equal to' in his equations. He chose the two lines because 'no two things can be more equal' (Cajori 1928, p. 126).

The etymology of the word 'equal' is from the Latin words 'aequalis' (meaning 'uniform', 'identical' or 'equal') and 'aequus' (meaning 'level', 'even' or 'just').

The symbol '=' was not immediately popular. The symbol '||' was used by some, and 'æ' (or 'œ'), from the Latin word 'aequalis' meaning 'equal', was widely used into the 1700s.

The History of the Equal Sign '=' in China

It seems there was no ancient symbol for '=' in Chinese, but the Chinese characters 等 *děng* (equality) for relational meaning and *dé* 得 (get the result) for procedural meaning were used broadly in ancient texts. Equality is related to the balance rule of *yin-yang* and the invariant principle of the *I Ching*. The basic procedures of *substituting* in the Chinese rod/*suàn pán*, *substituting 5 by 5 ones*, *substituting two 5s by 10*, *substituting 10 by 1 ten*, *substituting 100 by 10 tens*, *substituting 1 thousand by 10 hundreds,* etc., reflect the spirit of equality used in a broader, flexible way to some extent.

Such is the fundamental ancient Chinese mathematics spirit. 'Simultaneous equations' appears as one of the nine chapters of *The Nine Chapters on the Mathematical Art* (九章算术; *Jiǔzhāng Suànshù*) (Guo 2010). Spirit of equality is reflected in the 'equalising' and 'homogenising' theory (齐同原理), the first basic principle to deduce fractions, and 'cutting and paste' theory (割补原理), an explicit principle used when solving geometry problems involving area and volume in Liuhui's commentary on *The Nine Chapters* (Guo 2010).

There are 256 instances of the character 得 (*dé*) and 11 instances of the character 等 (*děng*) in *The Nine Chapters*. The fifth problem in 方田 *fangtian* – rectangular fields – reads as follows:

> The method for simplifying parts: What can be halved, halve them. As for what cannot be halved, separately set out the numbers for the denominator and numerator. Then alternately reduce them by subtraction. This is seeking for the *equality*. Simplify using this *equal* number. (Guo 2010, p. 99)

Fig. Comparing numbers in the first grade: the prior content of the textbook (Mathematics Textbook Developer Group for Elementary School 2005, p. 5)

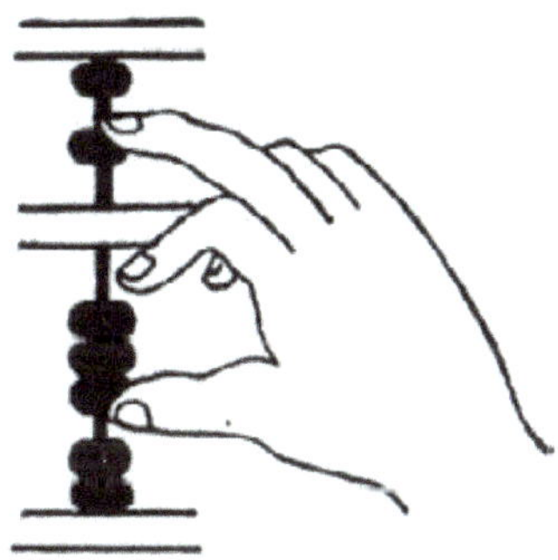

Fig. Fingering in *suàn pán*: the correct method of moving the beads (Kwa 1922, p. 8)

Chinese Approaches to the Relational Meaning of Equality

Ni (2015) reports that Chinese teachers are intolerant of errors where the relational (or conceptual) meaning of '=' is replaced by a procedural (or operational) meaning, while US teachers consider such errors minor. She mentions Chinese textbooks in which one-to-one correspondence is used from the beginning to assist students in better understanding the equal, greater-than and less-than symbols to enhance the relational meaning of '=' in contrast with '<' and '>' (Fig.).

This strategy is widespread in other countries (Alafaleq et al. 2015).

In general, the English expression 'how many' is translated into Chinese as 'more or less' (*duōshǎo*, 多少), which hints at the relational meaning and denotes a comparison of more than or less than (the imagined number). This expression is very common in arithmetic word problems. Like numerals, such expressions need a classifier highlighting the explicit connection between data and unknown values.

The variation approach to word problems presents another way to cope with the conceptual meaning of equality in China. Variation (变式, *biàn shì*) is a widely used approach that aims to discern the variance, invariance and sameness behind a group of problems and is regarded as the foundation of algebraic thinking and equations (Sun 2011, 2016). This approach is also closely related to the features of the Chinese language. Chinese is a tonal and logographic language, where each character has multiple meanings (一詞多義) and each word plays multiple roles in its context (一词多性). Teaching by variation is consistent with the needs of teaching the Chinese language. To learn to write Chinese and to increase their orthographic awareness, students must distinguish the similarities and differences of different characters that very often look similar to each other (Marton et al. 2010).

As variation problems enhance perceptions of variance and invariance or equality to solve word problems in Chinese curricula, they are regarded as one of the most important and explicit task design frameworks in China (Sun 2016). They refer to the 'routine' daily practice commonly accepted by Chinese teachers (Sun 2007, 2011; see also Cai and Nie 2007). Following Sun (2011), Bartolini Bussi et al. (2013, p. 550) describe a typical feature of these problems:

> One distinctive feature of word problems is to develop the ability to identify the invariant category of word problems (识类) it belongs to and discern different categories (归类), namely, discern the invariant elements from the variant elements between problems and recognize the 'class' every problem belong to. This pedagogy is generally called as *biànshì* (变式) in Chinese, where 'biàn' stands for 'changing' and 'shì' means 'form', can be translated loosely as 'variation' in English (Sun 2011). Some categories of biànshì are the following:
>
> OPMS (One Problem Multiple Solutions), where, for instance, the operation to solve the problem is carried out in different ways, with different grouping and ungrouping: $8 + 9 = (8 + 2) + 7$; $8 + 9 = 7 + (1 + 9)$ and so on.
>
> OPMC (One Problem Multiple Changes, see the variation problem below in Italy (Bartolini Bussi et al. 2013)), where in the same situation some changes are introduced.
>
> MPOS (Multiple Problem One Solution), where the same operation can be used to solve different problems, as in summary exercises (Sun 2011).

Western curricula use various models (e.g. models of taking away and comparing) to introduce meanings of addition/subtraction, as well as strategies to solve word problems. On the contrary, in Chinese curricula, rather than approaching word problems separately, problem variation permits them to be introduced in a holistic way without the use of multiple models (Sun 2015). Cai and Nie (2007, p. 467) report on the frequency of teaching with variation in the Chinese classroom through a survey of 102 teachers (see Table).

Table The frequency of teaching with variation in the Chinese classroom

	Used very often	Used occasionally	Never used
OPMS ($n = 102$)	84	18	0
OPMC ($n = 102$)	69	33	0
MPOS ($n = 100$)	52	48	0

An example of OPMS of addition with two digits is discussed in Chap. Section considers a transposition of additive variation problems to Italy.

Educational Implications

The above observations clearly point out some of the features of the Chinese arithmetic tradition:

- The inductive approach, where general principles of representing numbers and calculation are consistent with and derived from the specific case of the ones place (e.g. number operations in the tens/hundreds place are similar to number operations in the ones place).
- The tradition of calculation using specific cultural artefacts that also leave traces in the language.
- The variation tradition in word problems.

These features have important educational implications. Ma (1999) finds that the content knowledge of American and Chinese teachers is different. In particular, the strength of mathematics content knowledge is related to profound understanding of fundamental mathematics. According to Ma (1999):

> The US teachers tended to focus on the particular algorithm associated with an operation, for example, the algorithm for subtraction with regrouping, the algorithm for multi-digit multiplication, and the algorithm for division by fractions. The Chinese teachers, on the other hand, were more interested in the operations themselves and their relationships. In particular, they were interested in faster and easier ways to do a given computation, how the meaning of the four operations are connected, and how the meaning and the relationships of the operations are represented across subsets of numbers – whole numbers, fractions, and decimals. When they teach subtraction with decomposing a higher value unit, Chinese teacher start from addition with composing a higher value unit. (p. 112)

Similar reflections may be applied to other Western curricula. Chinese curricula do not have a chapter on place value similar to American or European curricula; rather, place value appears in all chapters, along with reading and writing number activities as an overarching principle. Place value involves implicit core knowledge of the number unit in ancient literature (Zou 2015) and in Chinese curricula (Sun 2015), which is different from the calculation vocabulary or extended number procedures in chapters on calculation in the mandatory practices of American curricula (Howe 2011, 2015).

Place Value and Whole Number Operations

Chinese verbal counting is transparent and completely regular for place value representation. However, in the West, place value may be perceived as an artificial construct for written purposes, as communities do not use it in ordinary conversation; for Western students, it can be a learned concept, but not a native one. The abbreviation of '-teen' numbers in English (13 to 19) and in other European languages cannot be easily decoded in terms of the place value of tens and ones, which hinders understanding of the ten-structured regroup aspects of a multi-digit calculation, i.e. addition with moving up a place/subtraction with moving back a place. This is consistent with the findings of Ho and Fuson (1998), who argue that the structure of the English language makes it more difficult to understand that '-teen' numbers are composed of a ten and some ones. It also makes it more difficult to learn the advanced make-a-ten method of single-digit addition and subtraction that is taught to first graders in China and other East Asian countries (Fuson and Kwon 1992; Geary et al. 1993; Murata 2004; Murata and Fuson 2001, 2006). Actually, the positional and decimal principles mentioned in Chapter (WG1) have been naturally embedded in Chinese numeration and everyday language since the third century BCE. Some scholars (e.g. Butterworth 1999) have interpreted this as a reason why Chinese students are at ease with place value for large numbers from the beginning. From ancient times until now, spoken Chinese whole numbers have been the same as written numbers, implying that the written numeral directly reflects its pronunciation and thus has not diverged from the spoken language. Place value is an unlearned activity, but it is an inherited concept like a mother language, where native speakers are often unaware of the complexities of their language. This may explain why all current Chinese curricula do not include the topic of place value.

Cardinal Numbers and Measure Numbers

From a conceptual perspective of numbers, Bass (see Chapter) points out that numbers and operations have two aspects: conceptual (what numbers are) and nominal (how we name and denote numbers). At least two possible pathways exist for the development of whole numbers: counting and measurement. Conceptually, numbers arise from a sense of quantity of some experiential species of objects: count (of a set or collection), distance, area, volume, time, rate, etc. To develop a conceptual understanding, Bass supports an approach to developing concepts of numbers using general notions of quantity and their measurement, in which the measurement 'unit' is key to knowing how much (or many) of the unit is needed to constitute the given quantity while measuring one quantity by another. Cardinal and measure numbers in Western languages appear very different from each other, as measure numbers require the choice of a unit. This is not the case in the Chinese language, where both are considered in the same way.

Fraction Names

The order of writing (and reading) a fraction in Western languages is 'first numerator, then denominator', and the denominator is usually named with ordinal (not cardinal) numbers, such as 'two thirds' (2/3) of three parts. (In Chinese naming, taking two parts indicates a part-whole relationship rather than 'two thirds'.) This method of fraction naming generates some difficulties for learning the part-whole relationship. The names for fractions in Western languages are not so clear. Bartolini Bussi et al. (2014) and Pimm and Sinclair (2015) analyse this difficulty and make proposals for overcoming it.

Arithmetic Operations

Research studies have identified several difficulties that Western students have with algorithms. For instance, Fuson and Li (2009) point out that many students in the USA make the error of subtracting the smaller number in a column from the larger number even if the smaller number is on the top:

$$\begin{array}{r} 346 \\ -\underline{157} \\ 211 \end{array}$$

This error may be reinforced by language confusion, as the names 'minuend' and 'subtrahend' do not emphasise the passive relationship between them (see above). However, this seems to be only a part of the story. Written algorithms for addition and subtraction were introduced in Europe by Leonardo Fibonacci in the thirteenth century. They hint at the actions performed on some kind of abacus (the Chinese *suàn pán*, the Japanese *soroban*, the Roman abacus or similar; see Menninger 1969). More recently, the spike abacus was introduced for teaching. In English and other Western languages, the operations in Figs. are described using terms like 'carrying'[6] and 'borrowing'.[7] The same was not true when algorithms were introduced in ancient textbooks. In *Liber abaci*, the term 'borrow' is not used. Rather, a kind of compensation or invariance is suggested: to increase by 10 the units in the minuend and to increase by 10 the units of the subtra-hend. In this process, the 10 to be added to the subtrahend must be 'kept in hands' (*reservare in manibus*, in Latin). This strategy was maintained in many method textbooks for primary schoolteachers in Italy until at least 1930.

[6] *Riporto - riportare* in Italian; *Übertrag* in German; *llevar* in Spanish; *retenue* in French.

[7] *Prestito* in Italian, *anleihe* in German, *prestar* in Spanish; *retenue* in French.

Ross and Pratt-Cotter (2000, 2008) reconstruct the story of the word 'borrowing' in North America. They find the first occurrence in a textbook by Osborne in 1827, but observe that 'the term borrow may be a misnomer since it suggests that something needs to be returned' (p. 49). Fuson and Li (2009) criticise this word (which was used for more than one century), and Fuson uses the words 'grouping' (for addition), 'ungrouping' (for subtraction) and 'regrouping' (if necessary in both cases) in the Math Expression project.[8]

The situation is quite different in China. In teaching subtraction with regrouping, the majority of the Chinese teachers interviewed in Ma (1999) describe the so-called 'borrowing' step in the algorithm as 'a process of decomposing a unit of higher value instead of saying "you borrow 1 ten from the tens place"' (p. 8). One third-grade teacher explained why she thought the expression 'decomposing a unit of higher value' was conceptually accurate:

> The term 'borrowing' can't explain why you can take 10 to the ones place. But 'decomposing' can. When you say decomposing it implies that the digits in higher places are actually composed of those at lower places. They are exchangeable. The term 'borrowing' does not mean the composing-decomposing process at all. (p. 9)

The English terms 'carrying' and 'borrowing' are not related to each other. The French term 'retenue' is the same for both operations. '*Retenue*' literally means 'keep in mind' (or 'keep in hand' in French) and hence hints at memory and not a concrete action. The origin may be traced back to the term used in medieval arithmetic '*reservare in manibus*' ('to keep in hands'). The use of the same term for different actions creates many difficulties for pupils (Soury-Lavergne, personal communication). This simple example shows that different cultures/languages may foster or hinder the understanding of meaning.

Mathematical Relational Thinking: Equality or Sameness

Some Reported Difficulties in the Understanding of Equality

Several studies have been carried out to examine the use of the equality symbol '=' in mathematics education. Kieran (1981) studies the interpretation of the equality symbol in the early grades. In preschool, two intuitive meanings appear: the first (*conceptual or relational meaning*) concerns the relation between two sets with the same cardinality (hence an equivalence relation, according to the historical genesis), while the second concerns the set resulting from the union of two sets. The second is related to the interpretation of '+' and '=' in terms of actions to be performed (*procedural or operational meaning*). This latter view is reinforced through the use

[8] http://www.hmhco.com/shop/education-curriculum/math/elementary-mathematics/math-expressions

of pocket calculators and the transcription of the additions and results as they appear on the display. For instance, to add the following numbers in a notebook:

$15 + 31 + 18$

it is common to see the following:

$15 + 31 = 46 + 18 = 64$

This discussion was carried out in a third-grade classroom in Italy. Only some excerpts are reported. The teacher (Rosa Santarelli) posed the following problem:

> How many days for holidays last summer?
> Two pupils have solved the problem as follows:
> 30 – 10 = 20 + 31 = 51 + 31 = 82 + 15 = 97
> Do you think that this calculation is correct?
> STE: Yes, it is correct. They have thought about the months of holidays. Hence, this month has so many days, and they have put that month. In June we were at school for 10 days, hence 30 – 10. … [T]hen they have written the equal sign and then 20 and from that 20 they have started to count all the holidays. They have written +31, then 51, + 31 equals 82, + 15 (the days in September) equals 97. Then they have understood the result, they have written it. What they have done is right.
> Many pupils agree and reword the same process.
> TEACHER: But what does the sign '='mean in mathematics?
> GIO: Equal means that if you have 20 + 30 you put the equal sign and you get the result. The equal sign tells the result of an operation …
> CAR: If you wish to use this sign in an operation, you must put it at the end. If you make 5 + 5 = then you write 10.
> Other pupils reword the same statements.
> TEACHER: What does it mean 'to be equal to' in mathematics?
> ILA: It means that you get the result.
> SAM: Equal, in mathematics, is usually in the operations. It is used to get the result.
> …
> TEACHER: Is it correct to write '8 = 8'?
> GIO: No, it isn't. You must write '+0' or else one doesn't understand. You need to put something.
> TEACHER: Hence, I make a mistake if I write '8 = 8'.
> GIO: Yes, you do. You should write '8 + 0 = 8' or '8 – 0 = 8'. (Zan 2007, p. 79 ff., our translation)

This short excerpt confirms that the procedural meaning of the equality symbol is often dominant in primary schools, at the expense of the relational meaning. Ni (2015) argues that student errors such as considering the equal sign as an order to 'do something' for an answer probably contribute to the difficulty they experience later when learning algebra; students treat an algebraic equation as indicating not a mathematical relation, but an order to 'do something' to obtain an answer. This may have very bad consequences in secondary school, when algebraic expressions are in the foreground. It is not possible to interpret the following equation according to the conceptual meaning:

$x + 3 = 4$

Teachers tacitly reinforce the procedural meaning when they do not take care of this issue.

Variation Problems in China and Italy

Bartolini Bussi et al. (2013) report an example of variation problems from the OPMC category, where all of the problems are collected in one 3 × 3 table (see also Sullivan et al. 2015, p. 88). In China, a collection of variation problems was given to second graders at the end of the school year as a kind of summary, with several different examples of problems presented during the school year. It was expected that the task would be solved in just one lesson due to the background knowledge of the students. Bartolini Bussi et al. used this task in some Italian schools, but a process of *cultural transposition* was needed.

The most evident effect of this transposition was the time needed. It was not possible to solve the task in just one lesson. The task was the source of a longer process, where the students had to become familiar with this surprising way of considering several problems together and using schemes to find/represent the solution. During the process, the students started to focus on the relationships between operations rather than on the execution of operations and hence started reasoning algebraically. Some further experiments (Mellone and Ramploud 2015) are in progress now.

Concluding Remarks

The attention to differences in whole number approaches is increasing. It is worthwhile to mention at least the book by Owens (2015), with a chapter on visuospatial reasoning with numbers, and the book by Owens et al. (2017) on the history of number in Papua New Guinea and Oceania that details number systems other than base 10 systems.

The examples discussed in this chapter show that language plays a common, key role in conveying concepts in the teaching and learning of whole number arithmetic. A cross-cultural examination of languages should thus allow us to understand linguistic supports and limitations that may foster or hinder students' learning and teachers' teaching of mathematics.

The above discussion highlights that in many cases the Chinese way to develop whole number arithmetic seems to offer advantages for the construction of mathematical meanings: the attention to mathematical consistency and coherence seems larger than in the Western curricula. Yet the Chinese case shows that the difference is strongly related to linguistic and cultural features not shared by other cultural groups. This observation suggests that caution must be taken when trying to apply some of the Chinese methods in other countries, unless a careful process of cultural transposition is established.

Table An example of variation problems from the OPMC category Beijing Education Science Research Institute and Beijing Instruction Research Center for Basic Education (1996), vol. 4, p. 88

Solve the following nine problems and then explain why they have been arranged in rows and columns in this way, commenting on their relationships.		
(1) *In the river there are 45* white ducks and 30 black ducks. How many ducks are there altogether?	(2) *In the river there are* white ducks and black ducks. There are 75 ducks altogether. 45 are white ducks. How many black ducks are there?	(3) *In the river there are* white ducks and black ducks. There are 75 ducks altogether. 30 are black ducks. How many white ducks are there?
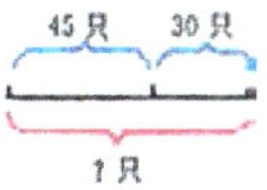	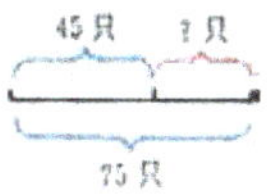	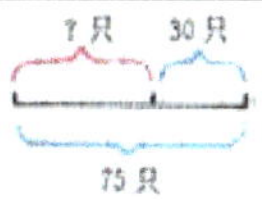
(1) *In the river there is* a group of ducks. 30 ducks swim away. 45 ducks are still there. How many ducks were in the group to begin with?	(2) *In the river there are* 75 ducks. Some ducks swim away. There are still 45 ducks. How many ducks swam away?	(3) *In the river there are* 75 ducks. 30 ducks swim away. How many ducks are still there?
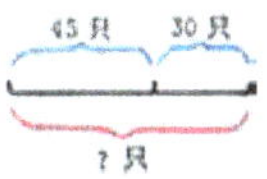	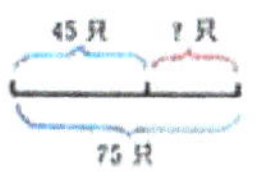	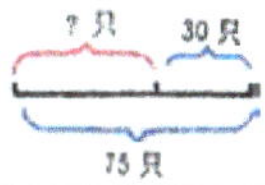
(1) *In the river there are* 30 black ducks. There are 15 more white ducks than black ducks (15 fewer black ducks than white ducks). How many white ducks are there?	(2) *In the river there are* 30 black ducks and 45 white ducks. How many more white ducks than black ducks (how many fewer black ducks than white ducks) are there?	(3) *In the river there are* 45 white ducks. There are 15 fewer black ducks than white ducks (15 more white ducks than black ducks). How many black ducks are there?
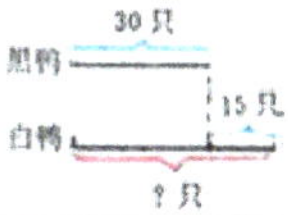	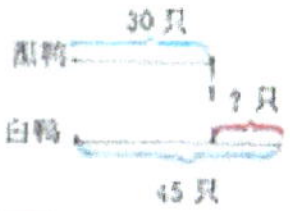	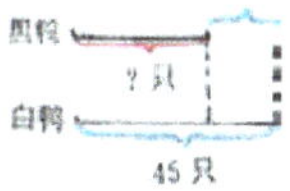

Acknowledgements These language and cultural issues were debated in different working groups, mainly working groups 1. We considered it important to collect contributions in a more systematic way in this chapter. We wish to thank all of the participants in the two working groups for providing interesting discussions and examples.

References

Allan, K. (1977). Classifiers. *Language, 53*(2), 285–311.

Askew, M. (2013). Big ideas in primary mathematics: Issues and directions. *Perspectives in Education, 31*(3), 5–18.

Bakker, M., van den Heuvel-Panhuizen, M., & Robitzsch, A. (2014). First-graders' knowledge of multiplicative reasoning before formal instruction in this domain. *Contemporary Educational Psychology, 39*, 59–73.

Bartolini Bussi, M. G. (2011). Artifacts and utilization schemes in mathematics teacher education: Place value in early childhood education. *Journal of Mathematics Teacher Education, 14*(2), 93–112.

Bartolini Bussi M. G., Sun X., & Ramploud A. (2013). A dialogue between cultures about task design for primary school. In C. Margolinas (Ed.), *Task design in mathematics education: Proceedings of ICMI study 22* (pp. 549–558). Retrieved 20 Jan 2016 from https://hal.archives-ouvertes.fr/hal-00834054

Bartolini Bussi, M. G., Baccaglini-Frank, A., & Ramploud, A. (2014). Intercultural dialogue and the history and geography of thought. *For the Learning of Mathematics, 34*(1), 31–33.

Barwell, R., Clarkson, P., Halai, A., Kazima, M., Moschkovich, J., Planas, N., Phakeng, M., Valero, P., & Villavicencio Ubillús, M. (Eds.). (2016). *Mathematics education and language diversity: The 21st ICMI study*. New York: Springer.

Bazzanella, C. (2011). Indeterminacy in dialogue. *Language and Dialogue, 1*(1), 21–43.

Butterworth, B. (1999). *The mathematical brain*. London: Macmillan.

Cai, J. (1998). An investigation of U.S. and Chinese students' mathematical problem posing and problem solving. *Mathematics Education Research Journal, 10*(1), 37–50.

Cai, J., & Knuth, E. (2011). *Early algebraization: A global dialogue from multiple perspectives*. Heidelberg: Springer.

Cai, J., & Nie, B. (2007). Problem solving in Chinese mathematics education: Research and practice. *ZDM Mathematics Education, 39*(5–6), 459–473.

Cajori, F. (1928). *A history of mathematical notations*. London: The Open Court Company.

Chauma, A. (2012). Teaching primary mathematical concepts in Chitumbuka: A quest for teacher education. *South African Journal of Higher Education, 26*(6), 1280–1295.

Chemla, K. (1996). Positions et changements en mathématiques à partir de textes chinois des dynasties Han à Song-Yuan. Quelques remarques. In Extrême-Orient, Extrême-Occident, n°18. Disposer pour dire, placer pour penser, situer pour agir: Pratiques de la position en Chine (pp. 115–147).

Chemla K. (1998). History of mathematics in China: A factor in world history and a source for new questions. *Doc. Math. J. DMV extra volume ICM III (1998)*, 789–798.

Cullen, C. (1996). *Astronomy and mathematics in ancient China: The zhou bi suan jing*. Cambridge: Cambridge University Press.

Dauben, J. (2002). Book review: The universal history of numbers and the universal history of computing (part 1). *Notices of the AMS, 49*(1), 32–38.

Fan, L., Wong, N. Y., Cai, J., & Li, S. (Eds.). (2004). *How Chinese learn mathematics: Perspectives from insiders*. Singapore: World Scientific.

Fan, L., Wong, N. Y., Cai, J., & Li, S. (2015). *How Chinese teach mathematics: Perspectives from insiders*. Singapore: World Scientific.

Funghi, S. (2016). Tables of numerals and others mathematical terms in different languages. Free download from http://www.umac.mo/fed/ICMI23/doc/tabels%20of%20numerals%20in%20different%20languages.pdf.

Fuson, K. C., & Briars, D. (1990). Using a base-ten blocks learning/teaching approach for first and second-grade place value and multidigit addition and subtraction. *Journal for Research in Mathematics Education, 21*(3), 180–206.

Fuson, K. C., & Kwon, Y. (1992). Korean children's understanding of multidigit addition and subtraction. *Child Development, 63*(2), 491–506.

Fuson, K. C., & Li, Y. (2009). Cross-cultural issues in linguistic, visual-quantitative, and written-numeric supports for mathematical thinking. *ZDM Mathematics Education, 41*(6), 793–808.

Galligan, L. (2001). Possible effects of English-Chinese language differences on the processing of mathematical text: A review. *Mathematics Education Research Journal, 13*(2), 112–132.

Geary, D. C., Bow-Thomas, C. C., Fan, L., & Siegler, R. S. (1993). Even before formal instruction, Chinese children outperform American children in mental addition. *Cognitive Development, 8*(4), 517–529.

Geary, D. C., Fan, L., & Bow-Thomas, C. C. (1992). Numerical cognition: Loci of ability differences comparing children from China and the United States. *Psychological Science, 3*(3), 180–185.

Guo, S. (2010). *Chinese history of science and technology*. Beijing: Science Press. [In Chinese].
Halai, A., & Karuka, S. (2013). Implementing language-in-education policy in multilingual mathematics classrooms: Pedagogical implications. *Eurasia Journal of Mathematics, Science & Technology Education, 9*(1), 23–32.
Ho, C. S. H., & Fuson, K. C. (1998). Children's knowledge of teen quantities as tens and ones: Comparisons of Chinese, British, and American kindergartners. *Journal of Educational Psychology, 90*(3), 536.
Howe, R. (2010). *Three pillars of first grade mathematics*. On the web at e.g: http://commoncoretools.me/2012/02/08/article-by-roger-howe-three-pillars-of-firstgrade- mathematics/
Howe, R. (2011). *The greatest calamity in the history of science*. Editorial, ICMI Newsletter 18.
Ifrah, G. (1981). *Histoire Universelle des Chiffres*. Paris: Editions Seghers/CNRS.
Kamii, C. (1986). Place value: An explanation of its difficulty and educational implications for the primary grades. *Journal of Research in Childhood Education, 1*(2), 75–86.
Kieran, C. (1981). Concepts associated with the equality symbol. *Educational Studies in Mathematics, 12*(3), 317–326.
Krifka, M. (2007). Approximate interpretation of number words: A case for strategic communication. In G. Bouma, I. Krämer, & J. Zwarts (Eds.), *Cognitive foundations of interpretation* (pp. 111–126). Amsterdam: Koninklijke Nederlandse Akademie van Wetenschapen.
Kwa, T. M. (1922). *The fundamental operations in bead arithmetic. How to use the Chinese abacus, Manila 1922*. Retrieved April 25, 2016, from http://www.google.it/url?sa=t&rct=j&q=&esrc=s&source=web&cd=1&ved=0ahUKEwj_pODQ2qnMAhXGWxoKHc2NDgUQFggdMAA&url=http%3A%2F%2Fpyramidal-foundational-information.com%2Ffile_download%2F415%2FChinese%2Babacus.pdf&usg=AFQjCNFC58zR0pYviS9QL9jU67WY0VO5mQ&sig2=ptok1mRe6gfMfug5bSmkBw
Lam, L. Y., & Ang, T. S. (1992/2004). *Fleeting footsteps: Tracing the conception of arithmetic and algebra in ancient China* (Revised edition). Singapore: World Scientific.
Lavric, E. (2010). Hyperbolic approximative numerals in cross-cultural comparison. In G. Kaltenböck, W. Mihatsch, & S. Schneider (Eds.), *New approaches to hedging* (pp. 123–164). Bingley: Emerald.
Leung F. K, Graf K. D., & Lopez-Real F. J. (Eds.) (2006). *Mathematics education in different cultural traditions – A comparative study of East Asia and the West: The 13th ICMI study*. New York: Springer.
Levin, S. R. (1982). Aristotle's theory of metaphor. *Philosophy & Rhetoric, 15*(1), 24–46.
Li, Y., & Huang, R. (Eds.). (2013). *How Chinese teach mathematics and improve teaching*. New York: Routledge.
Li, X., Ding, M., Capraro, M. M., & Capraro, R. M. (2008). Sources of differences in children's understandings of mathematical equality: Comparative analysis of teacher guides and student texts in China and the United States. *Cognition and Instruction, 26*(2), 195–217.
Ma, L. (1999). *Knowing and teaching elementary mathematics: Teachers' understanding of fundamental mathematics in China and the United States*. Mahwah: Lawrence Erlbaum Associates.
Marton, F., Tse, S. K., & Cheung, W. M. (2010). *On the learning of Chinese*. Rotterdam: Sense Publishers.
Martzloff, J. C. (1997). *A history of Chinese mathematics*. Berlin/Heidelberg: Springer Verlag.
Mathematics Textbook Developer Group for Elementary School. (2005). *Mathematics*. Beijing: People's Education Press. [In Chinese].
Menninger, K. (1969). *Number words and number symbols: A cultural history of numbers*. Cambridge, MA: The MIT Press. (Translated from the German edition of 1958).
Miller, K. F., Major, S. M., & Shu, H. (2000). Ordinal knowledge: Number names and number concepts in Chinese and English. *Canadian Journal of Experimental Psychology, 54*(2), 129–139.
Ministry of Education and Vocational Training. (2014). *Education and Training Policy*. (in Swahili). Retrieved from http://www.moe.go.tz/index.php?option=com_docman&Itemid=617.
Murata, A. (2004). Paths to learning ten-structured understanding of teen sums: Addition solution methods of Japanese grade 1 students. *Cognition and Instruction, 22*(2), 185–218.

Murata, A., & Fuson, K. (2001). Learning paths to 5- and 10-structured understanding of quantity: Addition and subtraction solution strategies of Japanese children. In R. Speiser, C. S. Maher, & C. Walter (Eds.), *Proceedings of the Twenty-third annual meeting of the North American Chapter of the International Group for the Psychology of Mathematics Education: Vol. 2* (pp. 639–646). Columbus: ERIC Clearinghouse for Science and Environmental Education: Mathematics.

Murata, A., & Fuson, K. (2006). Teaching as assisting individual constructive paths within an interdependent class learning zone: Japanese first graders learning to add using 10. *Journal for Research in Mathematics Education, 37*(5), 421–456.

National Curriculum Development Centre. (n.d.). *The national primary school curriculum for Uganda: Teachers' guide primary 1*. Kampala: The Author.

Needham, J., & Wang, L. (1959). *Mathematics and the sciences of the heavens and earth. The science and civilization in China series* (Vol. III). Cambridge: Cambridge Press.

Ng, S. S. N., & Rao, N. (2010). Chinese number words, culture, and mathematics learning. *Review of Educational Research, 80*(2), 180–206.

Owens, K. (2015). *Visuospatial reasoning: An ecocultural perspective for space, geometry and measurement education*. New York: Springer.

Owens, K., & Lean, G. with Paraide, P., & Muke, C. (2017). *History of number: Evidence from Papua New Guinea and Oceania*. New York: Springer.

Ran, Y. (2010). Approximative expressions and their loose uses in Chinese. In G. Kaltenb.ck, W. Mihatsch & S. Schneider (Eds.), *New approaches to hedging* (pp. 165–180). Bingley: Emerald.

Ross, S. H. (1989). Parts, wholes, and place value: A developmental view. *The Arithmetic Teacher, 36*(6), 47–51.

Ross, S., & Pratt-Cotter, M. (2000). Subtraction in the United States: An historical perspective. *The Mathematics Educator, 8*(1), 49–56.

Schwartzman, S. (1994). *The words of mathematics: An etymological dictionary of mathematical terms used in English*. Washington, DC: Mathematical Association of America.

Senft, G. (Ed.). (2000). *Systems of nominal classification*. Cambridge: Cambridge University Press.

Sepeng, P. (2014). Learners' voices on the issues of language use in learning mathematics. *International Journal of Educational Sciences, 6*(2), 207–215.

Siu, M. K. (2004). Official curriculum in mathematics in ancient China: How did candidates study for the examination? In L. Fan, N. Y. Wong, J. Cai, & S. Li (Eds.), *How Chinese learn mathematics: Perspectives from insiders* (pp. 157–185). Singapore: World Scientific.

Sun, X. (2007). *Spiral variation (*Bianshi*) curricula design in mathematics: Theory and practice*. Unpublished doctoral dissertation, Hong Kong: The Chinese University of Hong Kong [in Chinese].

Sun, X. (2011). Variation problems and their roles in the topic of fraction division in Chinese mathematics textbook examples. *Educational Studies in Mathematics, 76*(1), 65–85.

Sun, X. (2013). *The fundamental idea of mathematical tasks design in China: The origin and development*, paper presented in ICMI STUDY 22: Task design in mathematics education, University of Oxford, 22/07/2013–26/07/2013.

Sun, X. (2016). *Spiral variation: A hidden theory to interpret the logic to design Chinese mathematics curriculum and instruction in mainland China*. Singapore: World Scientific [in Chinese].

Sun, X. & Sun, Y. (2012). *The systematic model LǛ (率) of Jiu Zhang Suan Shu and its educational implication in fractional computation*, Paper presented at the 12th International Congress on Mathematics Education Seoul, Korea 08/07/2012–15/07/2012.

Sullivan, P., Knott, L., & Yang, Y. (2015). The relationships between task design, anticipated pedagogies, and student learning. In A. Watson & M. Ohtani (Eds.), *Task design: The 22nd ICMI study* (pp. 88–114). Cham: Springer.

Tempier, F. (2013). La numération décimale de position à l'école primaire: Une ingénierie didactique pour le développement d'une ressource. Thèse. Univ. Paris 7.

Verran, H. (2001). *Science and African logic*. Chicago: University of Chicago Press.
Volkov, A. (1996). Review of *Fleeting Footsteps*. *Archives Internationales d'Histoire des Sciences, 46*(136), 155–159.
Wang, J. (2013). *Mathematics education in China: Tradition and reality*. Singapore: CENGAGE Learning.
Wylie, A. (1853). Jottings on the sciences of the Chinese. *North Chinese Herald*, Shanghai.
Xie, X., & Carspecken, P. F. (2007). *Philosophy, learning and the mathematics curriculum: Dialectical materialism and pragmatism related to Chinese and American mathematics curriculums*. Rotterdam: Sense Publishers.
Zan, R. (2007). *Difficoltà in matematica: Osservare, interpretare, intervenire*. Milano: Springer.
Zaslavsky, C. (1973). *Africa counts: Number and pattern in African cultures*. Boston, MA: Prindle, Weber & Schmidt.

Cited Papers from Sun, X., Kaur, B., & Novotna, J. (Eds.). (2015). Conference proceedings of the ICMI study 23: Primary mathematics study on whole numbers. Retrieved February 10, 2016, from www.umac.mo/fed/ICMI23/doc/Proceedings_ICMI_STUDY_23_final.pdf

Alafaleq, M., Mailizar, L., Wang, Y., & Fan, L. (2015). How equality and inequality of whole numbers are introduced in China, Indonesia and Saudi Arabia primary mathematics textbooks (pp. 392–398).
Azrou, N. (2015). Spoken and written arithmetic in post-colonial countries: The case of Algeria (pp. 44–51).
Bass, H. (2015). Quantities, numbers, number name, and number line (pp. 10–20).
Ejersbo, L. R., & Misfeldt, M. (2015). The relationship between number names and number concepts (pp. 84–91).
Houdement, C., & Tempier, F. (2015). Teaching numeration units: Why, how and limits (pp. 99–106).
Howe, R. (2015). The most important thing for your child to learn about arithmetic (pp. 107–114).
Ma, L. (2015). The theoretical core of whole number arithmetic (pp. 34–38).
Mellone, M., & Ramploud, A. (2015). Additive structure: An educational experience of cultural transposition (pp. 567–574).
Ni, Y. (2015). How the Chinese methods produce arithmetic proficiency in children (pp. 339–345).
Pimm, D., & Sinclair, N. (2015). The ordinal and the fractional: Some remarks on a trans-linguistic study (pp. 354–361).
Sun, X. (2015). Chinese core tradition to whole number arithmetic (pp. 140–148).
Young-Loveridge, J., & Bicknell, B. (2015). Using multiplication and division contexts to build place-value understanding (pp. 379–386).
Zou, D. (2015). Whole number in ancient Chinese civilisation: A survey based on the system of counting-units and the expressions (pp. 157–164).

Permissions

All chapters in this book were first published in BFWNPG, by Springer Nature; hereby published with permission under the Creative Commons Attribution License or equivalent. Every chapter published in this book has been scrutinized by our experts. Their significance has been extensively debated. The topics covered herein carry significant findings which will fuel the growth of the discipline. They may even be implemented as practical applications or may be referred to as a beginning point for another development.

The contributors of this book come from diverse backgrounds, making this book a truly international effort. This book will bring forth new frontiers with its revolutionizing research information and detailed analysis of the nascent developments around the world.

We would like to thank all the contributing authors for lending their expertise to make the book truly unique. They have played a crucial role in the development of this book. Without their invaluable contributions this book wouldn't have been possible. They have made vital efforts to compile up to date information on the varied aspects of this subject to make this book a valuable addition to the collection of many professionals and students.

This book was conceptualized with the vision of imparting up-to-date information and advanced data in this field. To ensure the same, a matchless editorial board was set up. Every individual on the board went through rigorous rounds of assessment to prove their worth. After which they invested a large part of their time researching and compiling the most relevant data for our readers.

The editorial board has been involved in producing this book since its inception. They have spent rigorous hours researching and exploring the diverse topics which have resulted in the successful publishing of this book. They have passed on their knowledge of decades through this book. To expedite this challenging task, the publisher supported the team at every step. A small team of assistant editors was also appointed to further simplify the editing procedure and attain best results for the readers.

Apart from the editorial board, the designing team has also invested a significant amount of their time in understanding the subject and creating the most relevant covers. They scrutinized every image to scout for the most suitable representation of the subject and create an appropriate cover for the book.

The publishing team has been an ardent support to the editorial, designing and production team. Their endless efforts to recruit the best for this project, has resulted in the accomplishment of this book. They are a veteran in the field of academics and their pool of knowledge is as vast as their experience in printing. Their expertise and guidance has proved useful at every step. Their uncompromising quality standards have made this book an exceptional effort. Their encouragement from time to time has been an inspiration for everyone.

The publisher and the editorial board hope that this book will prove to be a valuable piece of knowledge for researchers, students, practitioners and scholars across the globe.

List of Contributors

Roger Howe
Yale University, New Haven, CT, USA

Claire Margolinas
Laboratoire ACTé, Université Clermont-Auvergne, Clermont-Ferrand, France

Xu Hua Sun
Faculty of Education, University of Macau, Macao, China

David Pimm
Simon Fraser University, Burnaby, BC, Canada

Christine Chambris
Laboratoire de didactique André Revuz, Université de Cergy-Pontoise, Cergy, France

Judy Sayers
Stockholm University, Stockholm, Sweden

Man Keung Siu
University of Hong Kong, Pokfulam, Hong Kong, China

Jason Cooper
Weizmann Institute of Science, Rehovot, Israel

Alf Coles
University of Bristol, Bristol, UK

Peter Gould
NSW Department of Education and Communities, Sydney, NSW, Australia

Jean-Luc Dorier
University of Geneva, Geneva, Switzerland

Sarah Inés González de Lora Sued
Pontificia Universidad Católica Madre, Santiago de los Caballeros, República Dominicana

Eva Thanheiser
Portland State University, Portland, OR, USA

Nadia Azrou
University Yahia Fares Medea, Medea, Algeria

Lynn McGarvey
University of Alberta, Edmonton, AB, Canada

Catherine Houdement
Laboratoire de didactique André Revuz, Université de Rouen-Normandie, Rouen, France

Lisser Rye Ejersbo
Aarhus University, Aarhus, Denmark

Joanne Mulligan
Macquarie University, Sydney, NSW, Australia

Lieven Verschaffel
Katholieke Universiteit Leuven, Leuven, Belgium

Anna Baccaglini-Frank
Università di Pisa, Pisa, Italy

Shengqing He
Hangzhou Normal University, Hangzhou, China

Yunpeng Ma, Yanling Wang and Shu Xie
Northeast Normal University, Changchun, China

Jasmina Milinković
University of Belgrade, Belgrade, Serbia

Andreas Obersteiner
Technische Universität München, München, Germany

Nicole Roberts
University of the Witwatersrand, Johannesburg, South Africa

Nathalie Sinclair
Simon Fraser University, Burnaby, BC, Canada

Der-Ching Yang
National Chiayi University, Chiayi, Taiwan

Bernard R. Hodgson
Université Laval, Quebec City, QC, Canada

Berinderjeet Kaur
National Institute of Education, Nanyang Technological University, Singapore, Singapore

Ann Gervasoni
Monash University, Melbourne, Victoria, Australia

Mike Askew
University of the Witwatersrand, Johannesburg, South Africa, and Monash University, Melbourne, Victoria, Australia

Michiel Veldhuis
Utrecht University, Utrecht, The Netherlands

Jarmila Novotná
Charles University, Prague, Czech Republic, and Laboratoire "Culture et Diffusion des Savoirs", Université de Bordeaux, Bordeaux, France

Catherine Pearn
The University of Melbourne, Melbourne, Victoria, Australia

Maria G. Bartolini Bussi
Department of Education and Humanities, University of Modena and Reggio Emilia, Modena, Italy

Index

www.ingramcontent.com/pod-product-compliance
Lightning Source LLC
LaVergne TN
LVHW051045220926
848485LV00001BA/15